T0073203

FUNCTIONAL INTERPRETATIONS

FUNCTIONAL INTERPRETATIONS

Justus Diller
University of Münster, Germany

 **World Scientific**

EW JERSEY · LONDON · SINGAPORE · BEIJING · SHANGHAI · HONG KONG · TAIPEI · CHENNAI · TOKYO

Published by

World Scientific Publishing Co. Pte. Ltd.
5 Toh Tuck Link, Singapore 596224
USA office: 27 Warren Street, Suite 401-402, Hackensack, NJ 07601
UK office: 57 Shelton Street, Covent Garden, London WC2H 9HE

Library of Congress Control Number: 2019044512

British Library Cataloguing-in-Publication Data
A catalogue record for this book is available from the British Library.

FUNCTIONAL INTERPRETATIONS
From the Dialectica interpretation to functional interpretations of analysis and set theory

ISBN 978-981-4551-39-7

For any available supplementary material, please visit
https://www.worldscientific.com/worldscibooks/10.1142/8945#t=suppl

This book is dedicated

to my wife Dorothea,
to my daughters Ricarda, Caroline, and Irene,
to my granddaughters Anna and Maike,
and to my colleague and friend Anne S. Troelstra
who died unexpectedly in March 2019.

Preface

The plan to write this book came up after a visit to California in early 2007, where I presented a functional interpretation of constructive set theory at UC Irvine on invitation by Kai F. Wehmeier and at UC Berkeley on invitation by John W. Addison. Large parts, the great majority of the LaTeX-files were expertly typed by Mrs. Hildegard Brunstering, in part even after her retirement from our institute. On the basis of these files, my colleague Anne S. Troelstra from the university of Amsterdam, suggested to add some introductory remarks to the single chapters which I think improve the readability of the text considerably. Later, my Münster colleague Wolfram Pohlers installed the LaTeX-system on my home computer and transferred the data concerning my manuscript to it. That was a decisive support for entering into the final stage of my manuscript which I eventually completed with the expert assistance of M. Sc. Jannes Bantje who was my consultant as well as my teacher in LaTeX in the final stage. I want to express my deep gratitude to these persons.

Münster, Germany, August 2019   Justus Diller

Contents

Chapter 0

Introduction

0.1 What is a functional interpretation?

On the occasion of the 70th birthday of the German logician Paul Bernays, then living in Zurich, Switzerland, in 1958, the Austrian logician Kurt Gödel, then living in Princeton, NJ, USA, published his paper "Über eine bisher nicht benützte Erweiterung des finiten Standpunktes" (*"On a hitherto unutilized extension of the finitary standpoint"*) [21] in the Swiss philosophical journal "Dialectica". In this paper of eight pages, Gödel outlines a functional interpretation of intuitionistic first order arithmetic, known as Heyting Arithmetic HA, in a theory T of primitive recursive functionals of finite types which came to be known as the Dialectica interpretation. By this interpretation, Gödel eliminates quantification from HA in favour of functionals in T. He thereby reduces the problem of consistency of HA to the problem of computability of the primitive recursive functionals. The concept of computability of finite type, he states, is to be considered as immediately comprehensible ("unmittelbar verständlich"), and it is this concept that he introduces as an extension of the finitary point of view mentioned in his title.

Gödel's Dialectica interpretation may be seen as expressing a lack of confidence or a distrust in our understanding of unbounded quantification. Instead of formally proving an implication with an existential consequent or with a universal antecedent, the Dialectica interpretation asks, under suitable conditions, for explicit "interpreting" instances that make the implication valid. Due to the decidability of the quantifier-free fragment of HA, and to the existence of characteristic terms for formulae in this fragment, such interpreting instances can be found. For proofs in extensions of HA with undecidable atomic formulae, or for proofs in constructive set theory, it may not always be possible to find just one such instance, but it must suffice to explicitly name a set consisting of interpreting instances - finitely many in the arithmetic case, set-many in the set-theoretic case. The aim of eliminating unbounded quantification in favor of appropriate functionals will still be obtained.

In the search for interpreting instances, the nesting of implications already in the language of HA calls for a theory of functionals of finite types.

0.1.1   *Inductive definition of finite types*

(1) o is a type.

(2) If σ, τ are types, then $(\sigma \to \tau)$ is a type.

(3) If σ, τ are types, then $(\sigma \times \tau)$ is a type.

The type o is the only **ground type**. Types of the form $(\sigma \to \tau)$ are **function types**, and types of the form $(\sigma \times \tau)$ are **product types**. The simplemost function type $o \to o$ is often denoted by 1. Types formed by rules (1) and (2) only, i.e. types not containing the symbol $\times$, are **linear types**. The set of linear types is denoted by $\mathcal{T}$, the set of all finite types by $\mathcal{T}^{\times}$. The word **type** refers ambiguously to linear types as well as to all finite types.

We omit round brackets when there is no danger of confusion. We use the convention of associating to the right and $\times$ binding stronger than $\to$.

Though the work in this book is mainly syntactical, we introduce the concept of type structure to give an idea of what types may stand for.

0.1.2   *Recursive definition of type structures*
$X^{\omega} = (X_{\tau} \mid \tau \in \mathcal{T}^{\times})$ *over a non-empty set X*

(1) $X_o = X$.

(2) $X_{\sigma \to \tau}$ is a set of operations ϕ which, to each element x of X_{σ}, assign a unique element $\phi(x)$ of X_{τ}, and is closed under explicit definitions.

(3) $X_{\sigma \times \tau}$ is the Cartesian product $X_{\sigma} \times X_{\tau}$ of X_{σ} and X_{τ}.

X^{ω} is the **full type structure** over X, if for all types σ, τ, $X_{\sigma \to \tau}$ is the class of all extensional functions from X_{σ} into X_{τ}.

There are more general definitions of type structure in the literature. For our purposes, however, the above defined concept suffices.

There is a nucleus of logical and combinatorial techniques which is common to all functional interpretations that we are going to study. It is essentially an intuitionistic propositional logic (without disjunction) of equations between typed terms and combinators which we may call a *basic logic of typed terms*, in short *BLTT*. Functional theories receive their strength from different recursors of finite type and corresponding induction schemata, yet they are all extensions of *BLTT* or of its type o version. We describe *BLTT* in detail and derive some elementary facts in this system.

0.1.3 *Description of a basic logic of typed terms BLTT and its type o version BLTT$_0$*

0.1.3.1 *Inductive definition of terms.* We write $a : \tau$ to indicate that a is a term of type τ.

(1) For any type τ, infinitely many **variables** $x^\tau, y^\tau, ...$ of type τ are terms of type τ.

(2) The constant 0 is a term of type o.

(3) For any types ρ, σ, τ, **combinators**
$\mathsf{K}_{\sigma\tau} : \tau \to \sigma \to \tau$ and $\mathsf{S}_{\rho\sigma\tau} : (\rho \to \sigma \to \tau) \to (\rho \to \sigma) \to (\rho \to \tau)$
are terms of the indicated types.

(4) In the presence of product types, for any types σ, τ, a
pairing functional $(\quad , \quad) : \sigma \to \tau \to \sigma \times \tau$ and
projection functionals $\mathsf{P}_0 : \sigma \times \tau \to \sigma$, $\mathsf{P}_1 : \sigma \times \tau \to \tau$
are terms of the indicated types.

(5) If $a : \sigma \to \tau$ and $b : \sigma$, then the **application** (ab) of a to b is a term of type τ.

Terms without variables are **functionals**, and atomic functionals are **constants**. So, in *linear* types, the only constants of *BLTT* are the constant $0 : o$ and the infinities of combinators K and S, but there will be further constants in functional theories extending *BLTT*. Again, we omit brackets when possible. Here, the convention is association to the left, i.e., $ab_1...b_n$ stands for $((...(ab_1)...)b_n)$. Furthermore, (a, b) stands for $(\quad , \quad)ab$.

0.1.3.2 *Inductive definition of formulae*

(1) **Equations of type** τ, i.e. equations $a = b$ between terms a, b, both of type τ, and the symbol $\perp$ (**falsehood**) are (atomic) formulae.

(2) If A and B are formulae, then their **conjunction** $(A \wedge B)$, and their **implication** $(A \to B)$ are formulae.

Negation $\neg A$ is defined as $(A \to \bot)$, **truth** $\top$ is $(\bot \to \bot)$, and **logical equivalence** $A \leftrightarrow B$ is $((A \to B) \land (B \to A))$. Formulae without (free) variables are **sentences**. Equations $a = b$ with a, b of a type $\tau \neq o$ are **equations of higher type**. Formulae in which there do not occur equations of higher type are **formulae of type** o. We omit brackets, following the same conventions as with types. So, $A \to B \to C$ stands for $(A \to (B \to C))$, and $A \land B \to C$ stands for $((A \land B) \to C)$. If in a formula F, all (free) occurrences of a variable x^τ are replaced by a term $a : \tau$, i.e. if a^τ is *substituted* for x^τ in F, we denote the resulting formula by $F_x[a]$. Usually, however, in this situation, when there is no danger of confusion, we shall write $F[x]$ for F and $F[a]$ for $F_x[a]$.

Like Gödel, we choose a system of "short" axioms and rules for $BLTT$, following essentially Spector [38], short in the sense that the formula schemata contain only few (at most 3) occurrences of metavariables for formulae - the metavariables themselves, of course, standing for arbitrarily long formulae. This will be of advantage when these schemata have to be functionally translated - a troublesome process for long formulae in any case.

0.1.3.3 Axioms and rules of inference of BLTT

A1	$A \to A$
A2	$A, A \to B \vdash B$
A3	$A \to B, B \to C \vdash A \to C$
A4	$A \land B \to A$
A5	$A \land B \to C \vdash A \to B \to C$
A6	$A \to B \to C \vdash A \land B \to C$
A7	$A \to B, A \to C \vdash A \to B \land C$
A$\bot$	$\bot \to A$
(K)	$\mathsf{K}_{\sigma\tau} a b = a$ for terms $a : \tau$, $b : \sigma$
(S)	$\mathsf{S}_{\rho\sigma\tau} a b c = ac(bc)$ for terms $a : \rho \to \sigma \to \tau$, $b : \rho \to \sigma$, $c : \rho$ which are equations of type τ.
(P)	$\mathsf{P}_0(a, b) = a$ and $\mathsf{P}_1(a, b) = b$ (in the presence of product types)
(Eq)	$A \to a = b, F[a] \vdash A \to F[b]$

Due to the lack of universal quantifiers in our language, the axioms of $BLTT$ are not necessarily sentences; they may be open formulae. We take care, however, that all axiom schemata are closed under substitution. This completes the description of the system $BLTT$.

0.1.3.4 *Description of* $BLTT_0$

Formulae of $BLTT_0$ are the formulae of type o of $BLTT$. With the exception of axioms (K), (S), and (P), **axioms and rules** of $BLTT_0$ are the same as for $BLTT$, restricted, however, to formulae of type o. If an equation $a = b$ of type τ is an axiom (K), (S), or (P) then all equations $c[a] = c[b]$ of type o, with an arbitrary context $c[z^\tau] : o$, are **axioms** of $BLTT_0$. Thus, also the description of the type o version $BLTT_0$ is completed.

We first collect a few elementary properties of the system $BLTT$.

0.1.4 *Lemma: Laws of implication and conjunction*

The following schemata and rules are derivable in $BLTT$:

(1) $A \to B \to A$
(2) $A \vdash B \to A$
(3) $A \wedge B \to B$
(4) $A \wedge B \to B \wedge A$
(5) $(A \wedge B) \wedge C \to A \wedge (B \wedge C)$
(6) $A \wedge (B \wedge C) \to (A \wedge B) \wedge C$
(7) $A \to B \to C \vdash B \to A \to C$
(8) $A \to B \vdash A \to A \wedge B$
(9) $A \to B, A \to B \to C \vdash A \to C$

Proof.

(1) follows from $A4$ by $A5$.
(2) follows from (1) by $A2$.
(3) follows by $A6$ from $A \to B \to B$ which follows by (2) from $A1$.
(4) follows by $A7$ from $A4$ and (3).
(5) $(A \wedge B) \wedge C \to D$ by $A4$ twice and $A3$ for $D \equiv A$, by $A4, (3)$, and $A3$ for $D \equiv B$, and by (3) for $D \equiv C$. Two applications of $A7$ yield (5).
(6) follows by reading the proof of (5) backwards, exchanging $A5$ and $A6$ as well as $A4$ and (3).
(7) by $A6$, (4), $A3$, and $A5$.
(8) by $A7$, applied to $A1$ and the premise.
(9) by (8), $A6$, and $A3$.

0.1.5   *Deduction Theorem*

*Let Th be an extension of BLTT by language, logic, and axioms - but without
any additional rules of inference. Then, for formulae D, A of $L(Th)$:*

$$If\ Th + D \vdash A\ ,\ then\ Th \vdash D \to A$$

Proof by induction on deductions. If A is an axiom of Th, then $D \to A$ by
0.1.4 (2). If $A \equiv D$, then $D \to A$ by $A1$.

$A2$. By induction hypothesis (I.H.) we have $D \to A$ and $D \to A \to B$; so
by 0.1.4 (9), $D \to B$ follows.

$A3$. By I.H., $D \to A \to B$ and $D \to B \to C$. By $A6$ and 0.1.4 (7), resp.,
we obtain $D \wedge A \to B$ and $B \to D \to C$. $A3$ now yields $D \wedge A \to D \to C$,
hence $D \wedge A \to C$ by $A4$ and 0.1.4 (9). From this follows $D \to A \to C$ by
$A5$.

$A5$. The I.H. $D \to A \wedge B \to C$ yields $D \wedge (A \wedge B) \to C$ by $A6$. From this,
$(D \wedge A) \wedge B \to C$ by 0.1.4 (5) and $A3$ from which, by two applications of
$A5$, the conclusion follows.

$A6$. The claim again follows by reading the argument for $A5$ backwards,
exchanging $A6$ and $A5$ as well as 0.1.4 (5) and (6).

$A7$. By I.H., we have $D \to A \to B$ and $D \to A \to C$ which by $A6$ and $A7$
yields $D \wedge A \to B \wedge C$ which by $A5$ implies the conclusion.

(Eq) By I.H. and $A6$, we have $D \wedge A \to a = b$ and $D \to F[a]$ which by (Eq)
give $D \wedge A \to D \to F[b]$. From this and $A4$, we get $D \wedge A \to F[b]$ which by
$A5$ implies the conclusion.

By induction on deductions, the deduction theorem now follows.

A word of caution: Whereas the rules in Lemma 0.1.4 are derived rules of
$BLTT$ that stay valid in any extension of $BLTT$ (within the language to
which the axioms and rules of inference pertain), the deduction theorem
states an admissible rule that may fail in extensions of $BLTT$ by additional
rules of inference.

0.1.6   *Corollary: Frege rule and Frege schema*

Frege rule and Frege schema are derivable in BLTT:

$$A \to B \to C \vdash (A \to B) \to A \to C$$

$$(A \to B \to C) \to (A \to B) \to A \to C$$

These follow by one resp. two applications of the deduction theorem to
lemma 0.1.4 (9). Again, Frege rule and Frege schema are hereby derived in
arbitrary extensions of $BLTT$.

0.1.7 Substitution Theorem for BLTT

If $\vdash F[x^\tau]$ and $a : \tau$, then $\vdash F[a]$

Since all axioms and rules of $BLTT$ are closed under substitution, the proof by induction on deductions is trivial.

There is just one rule (Eq) for equality. So, the first thing to verify about equality are the standard laws of *comparativity, reflexivity, symmetry, transitivity* and *replacement*.

0.1.8 Lemma: Equality laws

Standard laws of equality are derivable in BLTT:

$$
\begin{aligned}
&(comp) && a = b \to a = c \to b = c \\
&(refl) && a = a \\
&(sym) && a = b \to b = a \\
&(tran) && a = b \to b = c \to a = c \\
&(repl) && a = b \to c[a] = c[b]
\end{aligned}
$$

BLTT is an extension of $BLTT_0$.

Proof. In (Eq), put $A \equiv a = b$ and $F[x] \equiv a = c \to x = c$. We have $a = b \to a = b$ and $a = c \to a = c$ by $A1$, hence $(comp)$ by (Eq). By $(comp)$, we have $\mathsf{K}ab = a \to \mathsf{K}ab = a \to a = a$. By axiom (K) and $A2$, $(refl)$ follows. Identifying a and c in $(comp)$, we have $a = b \to a = a \to b = a$. By 0.1.4 (7), $(refl)$, and $A2$, (sym) follows. By $(comp)$ we have $b = a \to b = c \to a = c$. From (sym) and this, $(tran)$ follows by $A3$. For $(repl)$, in (Eq), put $A \equiv a = b$ and $F[x] \equiv c[a] = c[x]$. We then have $A \to a = b$ by $A1$ and $F[a]$ by $(refl)$, hence $A \to F[b]$ which is $(repl)$ by (Eq). By $(repl)$ and $A2$, all axioms of $BLTT_0$ are derivable in $BLTT$. Hence, $BLTT$ is a (derivability-)extension of $BLTT_0$.

Due to the presence of the constant $0 : o$, every type is inhabited:

0.1.9 Recursive definition of 0-functionals 0^τ of type τ

1. $0^o := 0$
2. $0^{\sigma \to \tau} := \mathsf{K}_{\sigma\tau} 0^\tau$
3. $0^{\sigma \times \tau} := (0^\sigma, 0^\tau)$

These 0-functionals satisfy $0^{\sigma \to \tau} a = 0^\tau$ for $a : \sigma$, $\mathsf{P}_0 0^{\sigma \times \tau} = 0^\sigma$, and $\mathsf{P}_1 0^{\sigma \times \tau} = 0^\tau$.

0.1.10   Recursive definition of λ-abstraction

We first define an **identity functional** $I_\tau : \tau \to \tau$ by

$$I_\tau :\equiv S_{\tau(\tau\to\tau)\tau} K_{(\tau\to\tau)\tau} K_{\tau\tau}$$

Omitting type indices, we obtain $Ib = SKKb = Kb(Kb) = b$ by axioms (S) and (K), hence by $(tran)$

$$Ib = b$$

Given types σ and τ, for any term $a : \tau$ and any variable x^σ, we define a λ-**abstraction term** $(\lambda x^\sigma.a)$ by recursion on subterms:

(λ1) $\lambda x^\sigma.x^\sigma :\equiv I_\sigma$
(λ2) $\lambda x^\sigma.a :\equiv K_{\sigma\tau} a$, if x^σ does not occur in $a : \tau$
(λ3) $\lambda x^\sigma.(ax^\sigma) :\equiv a$, if x^σ does not occur in $a : \sigma \to \tau$
(λ4) $\lambda x^\sigma.(ab) :\equiv S_{\sigma\rho\tau}(\lambda x^\sigma.a)(\lambda x^\sigma.b)$ otherwise for $a : \rho \to \tau, b : \rho$.

0.1.11   Lemma: Laws of λ-calculus

The expression $(\lambda x^\sigma.a)$ is, for $a : \tau$, a term of type $\sigma \to \tau$ in which the variable x^σ does not occur. Moreover, the λ-operator satisfies syntactic identities and in BLTT derivable equations:

(α) $\lambda x^\sigma.a[x^\sigma] \equiv \lambda y^\sigma.a[y^\sigma]$
(β) $(\lambda x^\sigma.a[x^\sigma])b = a[b]$ *for terms* $b : \sigma$
(η) $\lambda x^\sigma.(ax^\sigma) \equiv a$, *if* x^σ *does not occur in* a
(ξ') $b = c \to \lambda x^\sigma.a[x^\sigma, b] = \lambda x^\sigma.a[x^\sigma, c]$, *if* x^σ *does not occur in* b, c
(K') $\lambda x^\tau y^\sigma.x^\tau \equiv K_{\sigma\tau}$
(S') $\lambda x^{\rho\to\sigma\to\tau} y^{\rho\to\sigma} z^\rho.xz(yz) \equiv S_{\rho\sigma\tau}$

Proof. The initial statement and (α) are easily shown by running through definition 0.1.10. We check (β) by subterm induction:
(λ1) $(\lambda x.x)b = Ib = b \equiv x[b]$ as shown above.
(λ2) $(\lambda x.a)b = Kab = a \equiv a[b]$ by axiom (K), if x does not occur in a.
(λ3) $(\lambda x.ax)b = ab \equiv (ax)[b]$ if x does not occur in a.
(λ4) $(\lambda x.ac)b = S(\lambda x.a)(\lambda x.c)b = ((\lambda x.a)b)((\lambda x.c)b) = a[b]c[b] \equiv (ac)[b]$
by axiom (S) and I.H. otherwise.
Whereas (η) is simply $(\lambda 3)$, (ξ') holds, because by $(\lambda 2)$ and $(\lambda 3)$, maximal subterms that do not contain the abstraction variable x^σ are treated like atoms: The recursive definitions of $\lambda x^\sigma.a[x^\sigma, b]$ and $\lambda x^\sigma.a[x^\sigma, c]$ are simply the same, only exchanging occurrences of b and c.
$(K') : \lambda x^\tau y^\sigma.x^\tau \equiv \lambda x^\tau.K_{\sigma\tau}x^\tau \equiv K_{\sigma\tau}$ by $(\lambda 2)$ and $(\lambda 3)$

(S') : $\lambda x^{\rho\to\sigma\to\tau} y^{\rho\to\sigma} z^\rho.xz(yz) \equiv \lambda xy.\mathsf{S}_{\rho\sigma\tau}(\lambda z.xz)(\lambda z.yz) \equiv \lambda xy.\mathsf{S}_{\rho\sigma\tau}xy \equiv \lambda x.\mathsf{S}_{\rho\sigma\tau}x \equiv \mathsf{S}_{\rho\sigma\tau}$

By (β), $BLTT$ is **combinatorially complete**, i.e. it is closed under explicit definitions. Given a term $b[u^\tau]$, we may therefore say that a term f **is defined by** $fa = b[a]$ (for arbitrary $a : \tau$) instead of explicitly defining $f := \lambda u^\tau.b[u^\tau]$.

Up to and including definition 0.1.10 of the λ-operator, all proofs and results - restricted to formulae of type o - are literally correct also for $BLTT_0$. (α) and (η) in 0.1.10 are formal identities and hence independent of any deductive system. The statements $lb = b$ and (β) in lemma 0.1.11 should be read *under type o context* to become formally correct in $BLTT_0$: for $d : o$, $d[lb] = d[b]$ and

$$d[(\lambda x^\sigma.a[x^\sigma])b] = d[a[b]]$$

are proved as above. Law (ξ') may be handled the same way for $b, c : o$.

0.1.12 *Remark on iterated abstraction*

In spite of the fact that the variable x does not occur in $\lambda x.a[x]$, one must be careful when substituting into λ-terms. e.g.

$$(\lambda x.(\lambda y.a[x,y]))b = (\lambda y.a[b,y])$$

holds by (β) only in case y does not occur in b. Otherwise, the **substitutability** of b for x in $(\lambda y.a[x,y])$ is violated, and we replace y in $(\lambda y.a[x,y])$ by a new variable z not occurring in a, b which by (α) leaves the term unchanged, and obtain correctly by (β)

$$(\lambda x.(\lambda y.a[x,y]))b = (\lambda z.a[b,z])$$

When substituting into λ-terms, we shall from now on always assume substitutability.

As we shall frequently iterate processes like application and λ-abstraction, handle strings or *tuples* of terms, in particular of variables, we introduce a general convention.

0.1.13 *Convention on tuple notation*

1. If σ is a type tuple $\sigma_1, ..., \sigma_l$ and τ is a type tuple $\tau_1, ..., \tau_k$ then $\sigma \to \tau$ denotes the type tuple $\sigma_1 \to ... \to \sigma_l \to \tau_1, ..., \sigma_1 \to ... \to \sigma_l \to \tau_k$ (associating to the right). Hence, for $l = 0$, $\sigma \to \tau$ is τ, and for $k = 0$, $\sigma \to \tau$ is empty.

2. If a is a term tuple $a_1, ..., a_k$ of type tuple $\sigma \to \tau$ and b is a term tuple $b_1, ...,\ b_l$ of type tuple σ, then (ab) denotes the term tuple $a_1 b_1 ... b_l, ...,\ a_k b_1 ... b_l$ (associating to the left) of type tuple τ.

3. If x is a tuple $x_1, ..., x_l$ of variables of type tuple σ and $a \equiv a[x]$ is a term tuple $a_1, ..., a_k$ of type tuple τ then

$(\lambda x.a)$ and also $(\lambda x_1 ... x_l.a)$ denotes the term tuple

$$(\lambda x_1.\ ...(\lambda x_l.a_1)...),\ ...\ ,(\lambda x_1.\ ...(\lambda x_l.a_k)...)$$

of type tuple $\sigma \to \tau$. If this term tuple is applied to a term tuple $b : \sigma$, we assume that none of the variables from the tuple x occur in b. By induction on the length of the tuple σ, iterated application of (β) then yields

$$\vdash (\lambda x.a[x])b = a[b]$$

4. In the presence of quantifiers $\forall, \exists$, if x is a tuple of variables as above, and $A[x_1, ..., x_l]$ is a formula, then

$$\forall x A[x] \equiv \forall x_1 ... x_l A[x], \quad \exists x A[x] \equiv \exists x_1 ... x_l A[x]$$

stand respectively for the formulae

$$\forall x_1 ... \forall x_l A[x_1, ..., x_l] \text{ and } \exists x_1 ... \exists x_l A[x_1, ..., x_l].$$

5. If a, b are term tuples $a_i, b_i : \tau_i$ for $i = 1, ..., k$ of type tuple τ, then $a = b$ denotes the conjunction of equations $a_1 = b_1, ..., a_k = b_k$ of type tuple τ.

0.1.14   *Type-extensionaltity*

In λ-calculus, the equation (η) is known to lead, together with the ξ-rule, to extensionality. However, our law (ξ') is only a weak special case of this rule. Moreover, in its standard form, the

Axiom of type-extensionality

$(t - ext)$ $\forall x^\sigma f x^\sigma = g x^\sigma \to f = g$ for x^σ not free in $f, g : \sigma \to \tau$

makes formal use of an unbounded universal quantifier $\forall$ which is not available in $BLTT$. We therefore also look at *free parameter* formulations of these principles which are appropriate to extend $BLTT$ and its extensions:

Rule of type-extensionality

$(T - EXT)$ $A \to f x = g x \vdash A \to f = g$, if x is not free in A, f, g,

and the (ξ)-**rule**

(ξ) $A \to a = b \vdash A \to \lambda x^\sigma.a = \lambda x^\sigma.b$ for x not free in A.

Clearly, in systems with unbounded universal quantifier, the axiom schema

$(t-ext)$ and the rule $(T-EXT)$ are equivalent. But already in qf languages, we have:

0.1.14.1 Lemma. *On the basis of any extension of $BLTT$, the rules $(T-EXT)$ and (ξ) are equivalent.*

Proof. Given $(T-EXT)$ and $A \to a = b$ with x not free in A, but possibly in a, b, we have
$A \to (\lambda x.a)x = a = b = (\lambda x.b)x$ due to (β), hence by $(T-EXT)$
$A \to (\lambda x.a) = (\lambda x.b)$
Conversely, given (ξ) and $A \to fx = gx$ with x not free in A, f, g, we obtain
$A \to f = \lambda x.fx = \lambda x.gx = g$ by (η) and (ξ).

Extensions Th of $BLTT$ that satisfy $(T-EXT)$ for formulae A from a language L' smaller than $L(Th)$ are called **weakly extensional** or **WE-theories**, in contrast to **extensional theories** which satisfy the axiom $(t-ext)$. In extensional theories - which necessarily allow unbounded universal quantification -, $(T-EXT)$ is a derived rule, as already mentioned.

In languages with universal quantification, **extensional equality** $a =_e b$ (read: *a is extensionally equal to b*) is expressed by a type o formula:

0.1.14.2 *Recursive definition of* $a =_e b$ *for terms* $a, b : \tau$

(1) For $a, b : o$ $a =_e b :\equiv a = b$
(2) For $a, b : \sigma \to \tau$ $a =_e b :\equiv \forall x^\sigma \, a x^\sigma =_e b x^\sigma$ with a new variable x^σ
(3) For $a, b : \sigma \times \tau$ $a =_e b :\equiv \mathsf{P}_0 a =_e \mathsf{P}_0 b \wedge \mathsf{P}_1 a =_e \mathsf{P}_1 b$

The *tuple notation* extends to extensional equality in analogy to 0.1.13.5: If, in this definition, τ denotes a type-tuple $\tau_1, ..., \tau_k$, $a =_e b$ stands for the conjunction of $a_1 =_e b_1, ..., a_k =_e b_k$.

0.1.14.3 *Remark.* In the sequel, we restrict our attention to linear types. For these, clause (3) is never applied for the construction of $a =_e b$. For linear types $\tau \equiv \sigma \to o$ (σ a type tuple), definition 0.1.14.2 yields:

$a =_e b$ *is the type o formula* $\forall x \, ax = bx$ *with* $x : \sigma$ *a tuple of new variables, and the axiom of type-extensionality may equivalently be written in the form*

$(t-ext)$ $a =_e b \to a = b$

In type o languages, $(t-ext)$ is replaced by

$(t-ext_0)$ $a =_e b \to c[a] = c[b]$ for arbitrary contexts c of type o.

In extensions of $BLTT$ without universal quantifier, *positive occurrences* of $a =_e b$ are read without quantifiers, i.e. simply as $ax = bx$ with $x : \sigma$ a tuple of new variables. In type o languages, the rule $(T-EXT)$ is then replaced by

$(T - EXT_0)$ $A \to a =_e b \vdash A \to c[a] = c[b]$ for arbitrary contexts $c : o$
Here, the new variables that occur in $a =_e b$ must also not occur in the side
formula A.

The notion of type-extensionality may also be transferred from equations to
terms of (linear) type:

0.1.14.4 Type-extensional terms. A term $Y : \tau \equiv \sigma \to o$ (σ a type
tuple) is **extensional**, we write $E_\tau(Y)$, if

$$\forall u^\sigma \forall v^\sigma (u =_e v \to Yu = Yv)$$

Clearly, $(t - ext)$ implies $\forall Y \, E_\tau(Y)$.

For $\sigma = o, ..., o$, $u =_e v$ is just $u = v$, and $E_{\sigma \to o}(Y)$, in particular $E_1(Y)$ is
a standard law of equality. For later reference, we note the first non-trivial
case $E_2(Y)$ with $2 \equiv 1 \to o$:

$$E_2(Y) \equiv \forall u^1 \forall v^1 (\forall x^o (ux = vx) \to Yu = Yv)$$

and abbreviate the corresponding axiom $\forall Y \, E_2(Y)$ by (E_2).

Type-extensionality seems to be unavoidable in some extensions of $BLTT$
or of $BLTT_0$, e.g. in set theory in finite types, but it also complicates the
conceptual set-up of theories we shall work in. We therefore prefer to work
in *neutral* or *combinatorial* extensions of $BLTT$ as far as possible. Type-
extensionality may then be discussed as an additional, isolated ingredient.

We now come to the central definition of this introduction, the definition of
a general concept of functional interpretation of which, of course, Gödel's
1958 Dialectica interpretation [21] is the prototype.

0.1.15   *Definition of functional translations and functional interpretations*

A **functional theory** is an extension of $BLTT$ or of $BLTT_0$. A (mathematical) **theory** is either a constructive theory, containing unbounded
quantifiers $\exists u, \forall u$, with the variable u possibly of finite type, based on
intuitionistic predicate logic, or a classical theory in $\exists$**free formulation**, i.e.
with $\exists u \, A$ defined as $\neg \forall u \neg A$, based on classical predicate logic. A formula A
of such a theory Th is **qf** (or **quantifier-free**), if no unbounded quantifiers
$\exists u, \forall u$ occur in A (whereas bounded quantifiers $\forall x < t, \forall x \in t, \exists x \in t$ may
occur).

Let a theory Th and a functional theory FT be given such that the qf formulae of $L(Th)$ are formulae of $L(FT)$. A map I is a **functional translation** from $L(Th)$ into $L(FT)$, if, to any formula A of $L(Th)$, I recursively assigns an expression

$$A^I \equiv \exists v \forall w A_I[v, w]$$

where $A_I[v, w]$ is a formula of $L(FT)$ and v, w are (possibly empty) disjoint tuples of variables of finite type, also disjoint from $FV(A)$. A_I is called the **matrix**, $\exists v \forall w$ is called the **prefix** of A^I.

Moreover, a functional translation I will be required to be the identity on qf formulae, i.e.

$(qf)^I \quad A^I \equiv A_I \equiv A$ for quantifier free formulae A of $L(Th)$.

Functional translations studied in the sequel will also satisfy:

If A^I is as above and $B^I \equiv \exists y \forall z \, B_I[y, z]$, then
$(\wedge)^I \quad (A \wedge B)^I \equiv \exists v y \forall w z (A_I[v, w] \wedge B_I[y, z])$.

If $A[u]^I \equiv \exists v \forall w A_I[u, v, w]$, then
$(\forall)^I \quad (\forall u A[u])^I \equiv \exists V \forall u w A_I[u, Vu, w]$.

However, the ×-translations mentioned in the list 0.1.17 will not meet this requirement literally.

A functional translation I is a **functional interpretation** of Th in FT,

$$Th \overset{I}{\hookrightarrow} FT$$

if, for $Th \vdash A$, there is a tuple b of terms of $L(FT)$ (of the same type tuple as the variables v) such that

$$FT \vdash A_I[b, w]$$

and the variables in b are among the variables free in A. If this is the case for some term tuple b , A is called I-**interpretable in** FT, and the terms b are called (a tuple of) I-**interpreting terms** of A.

0.1.16   *Remarks*

(1) The recursive definition of A^I fixes the translation and even the interpretation of qf formulae, conjunctions, and (unbounded) universal quantifications for arbitrary I (cf. remarks (2),(3),(4) below), though, in the last case, it is not fixed whether the universal quantifier is added to the prefix or to the matrix of A^I (cf. remark (6) below). Disjunctions play only a subordinate role in this text.

As a primitive connective, disjunction $\lor$ occurs - like the bounded existential quantifier $\exists x \in t$ - only within Δ_0-formulae in set theory which count as qf formulae. So, different functional interpretations may differ essentially only in their translations of existential quantification and implication (cf. remark (7) below).

(2) In the translation $A^I \equiv A$ of qf A of $L(Th)$, the prefix is empty. Therefore, if a qf formula A is I-interpretable, it is I-interpreted by the empty tuple, and $FT \vdash A$. In such a case, we also say that A is I-**interpreted trivially**.

(3) If a I-interprets A and b I-interprets B, then, by $(\land)^I$, the tuple a, b I-interprets $A \land B$.

(4) In $(\forall)^I$, if v is a tuple of variables of type tuple σ and u is a single variable of type τ, then V is a tuple of variables of type tuple $\tau \to \sigma$, so that Vu is the tuple of terms $V_1 u, ..., V_n u$ of type tuple σ. And if the tuple $b[u]$ I-interprets $A[u]$, then by (β) in lemma 0.1.11, the tuple $\lambda u.b[u]$ I-interprets $\forall u A[u]$.

(5) In case the translated formulae A^I belong to $L(Th)$, we have by standard laws of (intuitionistic) logic

$$Th \vdash (A \land B)^I \leftrightarrow A^I \land B^I ,$$

but only in the presence of an appropriate choice principle do we get

$$Th \vdash (\forall u A[u])^I \leftrightarrow \forall u (A[u])^I .$$

The change of meaning of a formula under stepwise translation leads to the characterization problem discussed under 0.1.18.

(6) A functional translation I is a **modified realization** translation, if for any formula A, A^I is of the form $\exists v A_I[v]$, without universal quantifiers in the prefix. In that case, also $L(FT)$ contains unbounded universal quantifiers, and the quantifier $\forall u$ in $(\forall)^I$ becomes part of the matrix $(\forall u A)_I$ of $(\forall u A)^I$. In this situation, it is standard to say b I-**realizes** A and write b I A instead of $A_I[b]$.

(7) Functional translations I being recursively defined, $(A \to B)^I$ is constructed schematically out of the prefixes of A^I, B^I, substitution instances of A_I, B_I, and of A and B. Similarly, $(\exists u A[u])^I$ is constructed out of the prefix of $A[u]^I$ and substitution instances of $A[u]_I$ and $A[u]$. If the schematic construction of $(A \to B)^I, (\exists u A[u])^I$ makes use of data from A^I, B^I, resp. $A[u]^I$ only and does not mention the subformulae A, B, resp. substitution instances of $A[u]$, we

call I a **regular** functional translation. Otherwise, I is a **hybrid** translation.

(8) Even if A^I is a formula of FT, one should be aware of the special status of the existential part $\exists v$ of the prefix of A^I: If A is I-interpretable, then A^I, by definition, satisfies existential definability, since there are interpreting terms b with $FT \vdash A_I[b, w]$. If only $FT \vdash \exists v A[v]$, it is at least problematic whether there always is a b such that $FT \vdash A[b]$.

(9) If $Th \stackrel{I}{\hookrightarrow} FT$, the consistency of Th follows from the consistency of FT. In fact, Th is a conservative extension of FT, as far as qf formulae of $L(Th)$ are concerned.

(10) We do not study the concept of functional interpretations in full generality. The focus is rather on three regular translations and their hybrids. Among the three regular translations, the $\wedge$-translation (Diller-Nahm [15], Diller-Vogel [18]) in the arithmetic and analytic case, cf. sections 1.2 and 2.2, Burr-Schulte-Diller [6], [35], [13] in the set theoretic case, cf. section 3.3) is at the centre of attention, because the Dialectica translation turns out to yield interpretation theorems only for the type o fragment of the theories in question, in that case being equivalent to the $\wedge$-translation (cf. sections 1.3 and 3.3), and modified realization (cf. sections 1.6 and 3.4) is a weaker translation, leaving the unbounded universal quantifier unchanged.

We give a list of interpretation theorems for constructive as well as for classical theories that will be discussed in the sequel. Not all interpretations that appear in the list satisfy the above definition 0.1.15 of functional interpretation literally. However, these may easily be replaced by functional interpretations in the sense of 0.1.15, and that will be the form in which they will be treated. Also, Shoenfield [37] gives an interpretation of classical Peano arithmetic PA which he formulates in the $\{\neg, \vee, \forall\}$-fragment of predicate logic. His translation, in a way dual to the Dialectica translation, brings formulae of PA into a $\forall\exists$-form and violates definition 0.1.15 in several ways. However, if formulated in the $\{\bot, \wedge, \rightarrow, \forall\}$-fragment, the so-called *negative fragment* of predicate logic, PA is a subsystem of HA, and its Dialectica interpretation is just a corollary to Gödel's interpretation theorem. In the same way, the Diller-Nahm interpretation extends from HA^ω, the *natural span* of HA and Gödel's T, to its classical analogue PA^ω, the span of PA and T^c. Furthermore, Burr [7] may be read as a Diller-Nahm interpretation of Kripke-Platek set theory, if formulated in the negative fragment.

0.1.17 *List of interpretation theorems*

Gödel	1958 [21]:	$HA \overset{D}{\hookrightarrow} T_0$
		$PA \overset{D}{\hookrightarrow} T_0$
		$HA_0^\omega \overset{D}{\hookrightarrow} T_0$
		$I\!-\!HA^\omega \overset{D}{\hookrightarrow} T + E$
Kreisel	1959 [29]:	$HA^\omega \overset{mr}{\hookrightarrow} HA^\omega$
Spector	1962 [38]:	$WE\!-\!PA_0^\omega + CA_{0\tau} \overset{D}{\hookrightarrow} WE\!-\!T_0 + BR_\sigma$
Howard	1968 [23]:	$WE\!-\!HA_0^\omega + BI_\sigma \overset{D}{\hookrightarrow} WE\!-\!T_0 + BR_\sigma$
Troelstra	1971 [45]:	$HA^\omega \overset{mq}{\hookrightarrow} HA^\omega$
Diller-Nahm	1974 [15]:	$HA^\omega \overset{\wedge}{\hookrightarrow} T_\wedge$
		$PA^\omega \overset{\wedge}{\hookrightarrow} T_\wedge^c$
Stein	1974 [39]:	$HA^\omega \overset{\wedge q}{\hookrightarrow} HA^\omega$
Diller-Vogel	1975 [18] :	$HA^\omega + BI_\sigma \overset{\wedge}{\hookrightarrow} T_\wedge + SimBR_\sigma$
Stein	1980 [41]:	$HA^\omega \overset{n}{\hookrightarrow} HA^\omega$
Troelstra	1998 [48]:	$HA^\omega \overset{mrt}{\hookrightarrow} HA^\omega$
Burr	2000 [6]:	$CZF^{\omega-} \overset{\times}{\hookrightarrow} T_\in$
	2000 [7]:	$KP\omega^\omega \overset{\wedge}{\hookrightarrow} T_\in^c$
Schulte	2007 [35]:	$CZF^{\omega-} \overset{\times/\times q}{\hookrightarrow} T_\in/CZF^{\omega-}$
		$CZF^{\omega-} \overset{mr/mq/mrt}{\hookrightarrow} CZF^{\omega-}$
Diller	2008 [13]:	$CZF^{\omega-} \overset{\wedge/\wedge q}{\hookrightarrow} T_\in/CZF^{\omega-}$
Gaspar-Oliva	2011 [20]:	$HA^\omega \overset{\wedge t}{\hookrightarrow} HA^\omega$

Here, the functional interpretations $D, mr, \wedge, n,$ and $\times$ are regular, the others, $mq, \wedge q, mrt, \times q, \wedge t$ are hybrid translations (cf. remark 7 above).

0.1.18 *The characterization problem*

Once an interpretation theorem $Th \overset{I}{\hookrightarrow} FT$ is proved, one may ask how far proofs of some instance $A_I[b,w]$ of A^I in FT diverge from proofs of A in Th. If A^I is in $L(Th)$ for all formulae A of $L(Th)$ – which is the case with most functional interpretations mentioned above – , this amounts to characterizing the schema $\{A^I \leftrightarrow A\}$ in constructive theories and $\{A^{I-} \leftrightarrow A\}$ in classical theories, respectively, by axioms which do not mention the translation I. A^{I-} denotes the negative version $\neg\forall v \neg\forall w A_I[v,w]$ of $A^I \equiv \exists v \forall w A_I[v,w]$.

Like the interpretation theorems, the *characterization problem* has to be solved for each functional translation separately.

Definition, I-maximality, idempotent translations. Let I be a functional translation of a theory Th into a functional theory FT.
(1) We call Th I-**maximal relative to** FT, if every formula I-interpretable in FT is provable in Th.
(2) I is **idempotent**, if $A^I \in L(Th)$ and $A^{II} \equiv A^I$ for all $A \in L(Th)$.

I-maximality is a converse of I-interpretability. In the interpretation theorems $Th \overset{I}{\hookrightarrow} FT$ in the list above, the theories Th are in general not I-maximal relative to their interpreting functional theory FT. The only exceptions are the hybrid interpretations with truth mrt and $\wedge t$, simply because A^{mrt} as well as $A^{\wedge t}$ implies A in the respective theories. On the other hand, there does not even exist an I-maximal and I-interpretable theory (relative to FT), if the class of formulae I-interpretable in FT is not closed under deduction, as is the case with hybrid interpretations mq and $\wedge q$. However, if not only Th, but also the schema $\{A^I \leftrightarrow A\}$ is I-interpretable in FT, the situation changes:

0.1.19   *Lemma on I-maximality*

Let Th be an extension of FT.
(1) *If*

$$Th + \{A^I \leftrightarrow A\} \overset{I}{\hookrightarrow} FT$$

then $Th + \{A^I \leftrightarrow A\}$ is I-maximal relative to FT.
(2) *If I is a regular and idempotent functional translation of Th into FT, then*

$$Th \overset{I}{\hookrightarrow} FT \qquad implies \qquad Th + \{A^I \leftrightarrow A\} \overset{I}{\hookrightarrow} FT$$

and $Th + \{A^I \leftrightarrow A\}$ is again an I-maximal extension of Th relative to FT.

Proof. (1) Let B be I-interpretable in FT, i.e. there are terms b for which $FT \vdash B_I[b, z]$, the variables z not occurring in b. Then the extension Th of FT proves B^I, and $Th + \{A^I \leftrightarrow A\} \vdash B$. That proves (1).
(2) By regularity of I and by $(\wedge)^I$, $(A \leftrightarrow A^I)^I$ depends only on A^I and A^{II} which is again A^I, since I is idempotent, and is therefore identical with $(A \leftrightarrow A)^I$. So, $A \leftrightarrow A^I$ is I-interpretable in FT. That proves the first claim of (2), and the second follows by (1).

Chapter 1

Arithmetic

Though Gödel's Dialectica-interpretation is the first functional interpretation of Heyting arithmetic HA, in the first part of this chapter we put the $\wedge$-interpretation of Heyting arithmetic in finite types HA^ω in focus, because we consider the Dialectica-interpretation a special case of the $\wedge$-interpretation, working only for theories with atomic formulae equivalent to type o equations. After a presentation of the (essentially) quantifier free theories T and $T_\wedge$ of primitive recursive functionals of finite types, culminating in the theorem on generalized induction 1.1.22, the $\wedge$-interpretation of HA^ω is given in theorem 1.2.7 together with a characterization of the $\wedge$-translation in theorem 1.2.13. Dialectica-interpretation and -characterization of the type o fragment of HA^ω in the type o fragment of T as well as of the extension of HA^ω by equality functionals E_τ in $T + E_\tau$ then follow straight forward. All these results hold also in the corresponding theories in linear types only, due to the theorems on simultaneous recursions in section 1.4. To complete the discussion sketched in Gödel's Dialectica paper, strong computability of the primitive recursive functionals of finite type is proved in section 1.5 which in turn implies consistency of the theories considered here and continuity of the primitive recursive functionals.

In the following three sections, we turn to non-reductive functional interpretations, starting with a uniform treatment of modified realization and its q- and t-hybrids. Methods applied here are transferred to q- and t-hybrids of the $\wedge$-interpretation, leading to analogous results, though with considerable additional effort, in particular with respect to the axiomatic characterization of these hybrid translations. After that, an extension of T by a restricted universal set quantifier $\forall x \in a$ is introduced which makes possible a unified treatment of $\wedge$-interpretation and modified realization and, in between these two, of an infinity of so-called n-interpretations, for each type degree n.

Classical theories are formulated here generally in the negative fragment. Consequently, the $\wedge$- and Dialectica interpretations of theories like PA^ω are easy consequences of the corresponding results in the constructive case. Their characterization reduces in both cases to some special classical axioms of choice of which the interdependence, however, remains an open problem.

All theories functionally interpreted in this chapter avoid the axiom of extensionality $(t - ext)$. In the final section, the concept of majorizability is used to separate the primitive recursive functionals from functionals interpreting extensionality: Whereas primitive recursive functionals as well as equality functionals are majorizable, functionals that may D- or $\wedge$-interpret $(t - ext)$ are not.

1.1 Theories of primitive recursive functionals

We present Gödel's theory T of primitive recursive functionals of finite type as a theory in the negative fragment of intuitionistic propositional logic, i.e. without disjunction.

1.1.1 *Definition of Gödel's theory* T *as an extension of* $BLTT$ *and its type o version* T_0

The language of T extends the language of $BLTT$ only by the constants

$suc : o \to o$, the **successor function**, and, for every type τ,

the **recursor** $\mathsf{R}_\tau : (o \to \tau \to \tau) \to \tau \to o \to \tau$.

These are governed by the axioms

$$(Ax\ suc) \qquad suc\,t = 0 \to \bot$$
$$(Ax\ R0) \qquad \mathsf{R}_\tau ab0 = b$$
$$(Ax\ Rsuc) \qquad \mathsf{R}_\tau ab(suc\,t) = at(\mathsf{R}_\tau abt).$$

Axioms $(Ax\ R0)$ and $(Ax\ Rsuc)$ together form the **schema** (R) **of primitive recursion** in finite types.

Finally, the only rule added to $BLTT$ is the **rule of induction**

$$(IND) \qquad A[0],\ A[x^o] \to A[suc\,x^o] \vdash A[t]$$

for all formulae $A[x^o]$ of $L(T)$ and all terms $t : o$.

This completes the definition of in fact two versions of T, depending on whether we refer to all types, including product types, or just to linear types. The corresponding type o versions T_0 of either version of T are analogously defined as extensions of $BLTT_0$ by the same constants suc and

R_τ, by axiom $(Ax\ suc)$, by the axioms $(Ax\ R0)$ and $(Ax\ Rsuc)$ under type o context, and by the rule (IND) restricted to type o formulae. - Gödel's original T [21] has linear types only, contains a disjunction $\vee$ as an additional connective, and possibly, for each type τ, an additional equality functional $E_\tau : \tau \to \tau \to o$ which we will discuss in section 1.3.

First, we transfer logical properties from $BLTT$ to T:

1.1.2 Deduction Theorem for T

If a formula D of $L(T)$ does not contain variables of type o, then

$$T + \{D\} \vdash A \quad \text{implies} \quad T \vdash D \to A.$$

Proof by induction on deductions. By the deduction theorem 0.1.5 for $BLTT$, the only induction step to be considered here is the rule (IND). If, by I.H., $D \to A[0]$ and $D \to A[x] \to A[suc\,x]$ are derivable in T, then so is $(D \to A[x]) \to (D \to A[suc\,x])$ due to the Frege rule 0.1.6. Only in case the eigenvariable x^o does not occur in D, $T \vdash D \to A[t]$ follows by (IND). That is why we exclude variables of type o from the deduction formula D.

1.1.3 Substitution Theorem for T

If $\vdash F[u^\tau]$ and $a : \tau$, then $\vdash F[a]$.

Proof. As $Ax(T)$ is closed under substitutions, again only the rule (IND) has to be considered. If $T \vdash A[x, u] \to A[suc\,x, u]$, we may, by I.H., change the eigenvariable x to a variable y^o which does not occur in A, b. Again by I.H., we have $T \vdash A[y, b] \to A[suc\,y, b]$, and (IND) now gives the desired result $T \vdash A[t, b]$.

The rule of induction (IND) contains, as a special case, a rule of proof by cases:

1.1.4 Lemma: Proof by cases

Proof by cases is a derived rule in T:
$(Cases) \qquad A[0],\ A[suc\,x^o] \vdash A[t]$

This follows from (IND) by lemma 0.1.4 (2).

At first sight, the axioms for the recursor look like the equations for primitive recursion in elementary recursion theory. However, since the axioms for R_τ are equations of type τ, their expressive power is far superior to elementary primitive recursion. It is an exercise to show that the Ackermann function is definable in T by an application of $\mathsf{R}_{o\to o}$.

A look at $(Ax\ Rsuc)$ shows that the recursor contains three special or degenerate cases:

(1) The term a may be independent of its first argument t; in this case, recursion becomes iteration.

(2) a may be independent of its second argument $\mathsf{R}_\tau abt$; in this case, recursion becomes case distinction, defining a function of type $o \to \tau$ outright as b at point $t = 0$ and as at at all points $suc\ t$.

(3) Or a may be independent of both arguments in which case recursion degenerates to case distinction between just two values:

1.1.5 Definition and Lemma: Iteration and case distinction

For any type τ, we define an **iterator**

$$J_\tau : (\tau \to \tau) \to \tau \to o \to \tau \qquad \text{by} \quad J_\tau a = \mathsf{R}_\tau(Ka)$$

and **case distinction functionals**

$$C_\tau : (o \to \tau) \to \tau \to o \to \tau \qquad \text{by} \quad C_\tau a = \mathsf{R}_\tau(\mathsf{S}(\mathsf{KK})a)$$

$$D_\tau : \tau \to \tau \to o \to \tau \qquad \text{by} \quad D_\tau a = J_\tau(Ka) = \mathsf{R}_\tau(\mathsf{K}(Ka))$$

These functionals satisfy

$$
\begin{array}{lll}
Jab0 = b & \text{and} & Jab(suc\,t) = a(Jabt) \\
Cab0 = b & \text{and} & Cab(suc\,t) = at \\
Dab0 = b & \text{and} & Dab(suc\,t) = a
\end{array}
$$

Proof. $Jab0 = Cab0 = Dab0 = \mathsf{R}(...)b0 = b$,

$Jab(suc\,t) = \mathsf{R}(Ka)b(suc\,t) = Kat(\mathsf{R}(Ka)bt) = a(Jabt)$,

$Cab(suc\,t) = \mathsf{R}(\mathsf{S}(\mathsf{KK})a)b(suc\,t) = \mathsf{S}(\mathsf{KK})at(\mathsf{R}(\mathsf{S}(\mathsf{KK})a)bt)$

$\qquad\qquad\quad = \mathsf{KK}t(at)(Cabt) = \mathsf{K}(at)(Cabt) = at$,

$Dab(suc\,t) = J(Ka)b(suc\,t) = Ka(J(Ka)bt) = a$.

These calculations are quite straight forward, and we shall not worry to go into such detail later on.

First applications of these functionals are addition as iteration of the successor function, multiplication as iteration of addition (of the first argument), predecessor as case distinction by C_o between 0 and identity I_o, arithmetic difference as iteration of the predecessor, and the sign functions. We list explicit definitions of some standard primitive recursive functions and relations that we make use of in the sequel:

1.1.6 *List of definitions of primitive recursive functions and relations*

addition	$+ : o \to o \to o$	by $+ = J_o suc$
	and write	$(s + t)$ for $+ st$
multiplication	$\cdot : o \to o \to o$	by $\cdot s = J_o(+s)0$
	and write	$(s \cdot t)$ for $\cdot st$
finite sum	$\Sigma : 1 \to o \to o$	by $\Sigma f = R_o(\lambda xz.z + (fx))0$
	and write also	$\Sigma_{x<t} fx$ for Σft
finite product	$\Pi : 1 \to o \to o$	by $\Pi f = R_o(\lambda xz.z \cdot (fx))1$
	and write also	$\Pi_{x<t} fx$ for Πft
predecessor	$prd : o \to o$	by $prd = C_o \mathsf{I}_o 0$
arithmetic difference	$\dot- : o \to o \to o$	by $\dot- = J_o prd$
	and write	$(s \dot- t)$ for $\dot- st$
absolute difference	$\mid - \mid : o \to o \to o$	by $\mid - \mid st = (s \dot- t) + (t \dot- s)$
	and write	$\mid s - t \mid$ for $\mid - \mid st$
sign function	$sg : o \to o$	by $sg = D_o 10$
equality function	$E_o : o \to o \to o$	by $E_o st = sg \mid s - t \mid$
less-or-equal relation	$\leq$	by $s \leq t :\equiv s \dot- t = 0$
less-than relation	$<$	by $s < t :\equiv suc \, s \dot- t = 0$
maximum (of two)	$max : o \to o \to o$	by $max \, st = (s + (t \dot- s))$
iterated maximum	$Max : 1 \to o \to o$	by $Max \, f = R_o(\lambda xz.max \, z(fx))0$
	and write also	$max\{fx \mid x < t\}$ for $Max \, ft$
bounded μ-operator	$\mu : 1 \to o \to o$	by $\mu f = J_o(\lambda z.z + sg(fz))0$
	and write also	$\mu x < t(fx = 0)$ for μft
ordered pair	$\langle , \rangle : o \to o \to o$	by $\langle s, t \rangle = (\Sigma \mathsf{I}_o(suc(s+t))) + s$
left decoding	$j_0 : o \to o$	by $j_0 r = r \dot- \langle 0, a[r] \rangle$
right decoding	$j_1 : o \to o$	by $j_1 r = a[r] \dot- j_0 r$
	with	$a[r] = \mu z < suc \, r(r < \langle 0, suc \, z \rangle)$

We now give a fairly long list of standard elementary properties of these primitive recursive functions and relations. The reader should verify that these properties, if arranged in proper order, have simple quantifier-free proofs in T.

1.1.7 *Lemma: Laws of successor, predecessor, sign, and case distinction*

The following laws of suc, prd, sg, D_o are derivable in T:

$(prd0)$ $prd\,0 = 0$
$(prd1)$ $prd(suc\,t) = t$
$(prd2)$ $t \neq 0 \to t = suc(prd\,t)$
$(suc1)$ $suc\,s = suc\,t \to s = t$
$(D1)$ $D_o 100 = sg0 = 0$
$(D2)$ $D_o 10(suc\,t) = sg(suc\,t) = 1$
$(D3)$ $D_o 01t = 1 \dot{-} t$

Proof. $(prd0)$ and $(prd1)$ are the recursion equations for C in lemma 1.1.5 with $a = \mathsf{I}$. $(prd2)$ follows by induction on t. $(suc1)$ follows from $(prd1)$. $(D1)$ to $(D3)$ are immediate from the recursion equations for D in 1.1.5 and for $\dot{-} = J_0 prd$.

1.1.8   Lemma: Laws of addition

The following laws of addition $+$ are derivable in T:
$(+0)$ $s + 0 = s$
$(+1)$ $s + suc\,t = suc(s + t)$
$(+2)$ $r + (s + t) = (r + s) + t$
$(+3)$ $0 + t = t$
$(+4)$ $suc\,s + t = suc(s + t)$
$(+5)$ $s + t = t + s$
$(+6)$ $r + t = s + t \to r = s$
$(+7)$ $s + t = 0 \to s = 0 \wedge t = 0$

Proof. $(+0)$ and $(+1)$ are the recursion equations for addition. $(+2)$ through $(+5)$ follow by induction on t, with $(+5)$ making use of $(+3)$ and $(+4)$. $(+6)$ is proved by induction on t and $(suc1)$, $(+7)$ by induction on t and $(Ax\ suc)$.

1.1.9   Lemma: Laws of multiplication

The following laws of multiplication $\cdot$ are derivable in T:
$(\cdot0)$ $s \cdot 0 = 0$
$(\cdot1)$ $s \cdot suc\,t = s + s \cdot t$
$(\cdot2)$ $r \cdot (s + t) = r \cdot s + r \cdot t$
$(\cdot3)$ $r \cdot (s \cdot t) = (r \cdot s) \cdot t$
$(\cdot4)$ $0 \cdot t = 0$
$(\cdot5)$ $suc\,s \cdot t = t + s \cdot t$
$(\cdot6)$ $s \cdot t = t \cdot s$

Proof. ($\cdot 0$) and ($\cdot 1$) are the recursion equations for $\cdot$. Due to our short definition of $\cdot s$, ($\cdot 1$) comes out in an unusual order which, however, in the presence of the commutativity of addition, ($+5$), does not matter. ($\cdot 2$) to ($\cdot 6$) follow – in this order – by induction on t. We show ($\cdot 5$): We have
$suc\, s \cdot 0 = 0 = 0 + s \cdot 0$ and
$$suc\, s \cdot suc\, x = suc\, s + suc\, s \cdot x = suc\, s + (x + s \cdot x) \quad \text{by } (\cdot 1) \text{ and I.H.}$$
$$= suc\, x + (s + s \cdot x) \quad \text{by } (+4), (+2), (+5), \text{ and } (+4)$$
$$= suc\, x + s \cdot suc\, x) \quad \text{by } (\cdot 1), \text{ and } (IND) \text{ now proves } (\cdot 5).$$

1.1.10 Lemma: Laws of finite sum and finite product

Recursion equations for finite sum and finite product in T:
($\Sigma 0$) $\Sigma f 0 = 0$
($\Sigma 1$) $\Sigma f(suc\, t) = (\Sigma ft) + ft$
($\Pi 0$) $\Pi f 0 = 1$
($\Pi 1$) $\Pi f(suc\, t) = (\Pi ft) \cdot ft$

1.1.11 Lemma: Laws of arithmetic difference

The following properties of arithmetic difference $\dot{-}$ are derivable in T:
($\dot{-}0$) $s \dot{-} 0 = s$
($\dot{-}1$) $s \dot{-} suc\, t = prd(s \dot{-} t)$
($\dot{-}2$) $0 \dot{-} t = 0$
($\dot{-}3$) $suc\, s \dot{-} suc\, t = s \dot{-} t$
($\dot{-}4$) $r \dot{-} (s + t) = (r \dot{-} s) \dot{-} t$
($\dot{-}5$) $(s + t) \dot{-} t = s$
($\dot{-}6$) $t \dot{-} t = 0$, *or shorter* $t \leq t$
($\dot{-}7$) $suc\, s \dot{-} t = 0 \to s \dot{-} t = 0$, *or shorter* $s < t \to s \leq t$

Proof. ($\dot{-}0$) and ($\dot{-}1$) are the recursion equations for $J_o prd$. ($\dot{-}2$) to ($\dot{-}4$) follow by induction on t. ($\dot{-}5$) to ($\dot{-}7$) follow from ($\dot{-}3$) by induction on t.

1.1.12 Proposition: Induction along the diagonal

In T, the following rule of induction along the diagonal is derivable:
$$A[x, 0], \ A[0, y], \ A[x, y] \to A[suc\, x, suc\, y] \ \vdash \ A[s, t]$$

Proof. 1. $A[x, prd\, 0] \to A[suc\, x, 0]$ follows by 0.1.4 (2) from the first premise,
2. $A[x, prd(suc\, y)] \to A[suc\, x, suc\, y]$ follows by $(prd1)$ in 1.1.7 from the third premise, hence
3. $A[x, prd\, y] \to A[suc\, x, y]$ by $(Cases)$ 1.1.4. Similarly, we get

4. $A[prd\,x, prd\,y] \to A[x,y]$ by $(Cases)$

from the second premise and 3. This becomes

5. $A[s \dot{-} suc\,z, t \dot{-} suc\,z] \to A[s \dot{-} z, t \dot{-} z]$ by $(\dot{-}1)$.

By Frege's rule 0.1.6, 5. implies

6. $(A[s \dot{-} z, t \dot{-} z] \to A[s,t]) \to (A[s \dot{-} suc\,z, t \dot{-} suc\,z] \to A[s,t])$. Since

7. $A[s \dot{-} 0, t \dot{-} 0] \to A[s,t]$ by $(\dot{-}0)$ and (Eq), we have

8. $A[s \dot{-} t, t \dot{-} t] \to A[s,t]$ by (IND), applied to 7. and 6.

Since $t \dot{-} t = 0$ by $(\dot{-}6)$,

9. $A[s,t]$ now follows from the first premise and 8. by *modus ponens* A2.

$A[x,y]$ is occasionally called the *induction hypothesis* (I.H.) of this rule.

Induction along the diagonal simplifies some proofs considerably:

1.1.13   Lemma

In T *is derivable*

$(\dot{-}8)$	$t \dot{-} s = suc\,r \to suc\,s \dot{-} t = 0$, i.e. $\neg t \leq s \to s < t$
$(\dot{-}9)$	$t \dot{-} s = 0 \wedge s \dot{-} t = 0 \to s = t$, i.e. $t \leq s \wedge s \leq t \to s = t$
$(|-|1)$	$|s - t| = 0 \to s = t$
$(|-|2)$	$E_0 st = 0 \leftrightarrow |s - t| = 0 \leftrightarrow s = t$
$stab(=:o)$	$\neg\neg s = t \to s = t$
$(max1)$	$max\,st = max\,ts$
$(max2)$	$s \leq t \leftrightarrow max\,st = t$
$(\leq trans)$	$r \leq s \wedge s \leq t \to r \leq t$
$(Max0)$	$Max\,f0 = 0$
$(Max1)$	$Max\,f(suc\,t) = max(Max\,ft)(ft)$
$(Max2)$	$s < t \to fs \leq Max\,ft$

Proof. $(\dot{-}8)$: $0 \dot{-} x = 0 = suc\,r$ implies $\bot$ and hence $suc\,x \dot{-} 0 = 0$. And $y \dot{-} 0 = y = suc\,r$ by $(\dot{-}3)$ implies $suc\,0 \dot{-} y = 0 \dot{-} r = 0$. Now, again by $(\dot{-}3)$, the I.H. implies

$$suc\,y \dot{-} suc\,x = y \dot{-} x = suc\,r \to suc\,suc\,x \dot{-} suc\,y = suc\,x \dot{-} y = 0.$$

Proposition 1.1.12 now yields $(\dot{-}8)$.

$(\dot{-}9)$: $0 \dot{-} x = 0 \wedge x \dot{-} 0 = 0 \to x = 0$ and $y \dot{-} 0 = 0 \wedge 0 \dot{-} y = 0 \to y = 0$. By $(\dot{-}3)$ and the induction hypothesis

$$suc\,y \dot{-} suc\,x = y \dot{-} x = 0 \wedge suc\,x \dot{-} suc\,y = x \dot{-} y = 0$$

implies $x = y$ and hence $suc\,x = suc\,y$. 1.1.12 now yields $(\dot{-}9)$.

In view of $(+7)$, $(\dot{-}9)$ implies $(|-|1)$ which, with $(\dot{-}6)$, implies $(|-|2)$.

$stab(=:o)$: $\neg\neg 0 = 0 \to 0 = 0$ and, since $\neg\neg suc\,x = 0 \to \bot$, also

$$\neg\neg suc\,x = 0 \to suc\,x = 0;$$

so, by (IND) we have

$$\neg\neg|s - t| = 0 \to |s - t| = 0$$

which by $(| - |2)$, implies $stab(= : o)$.

$(max1)$: $x + (0 \dot{-} x) = x = 0 + (x \dot{-} 0)$ and $0 + (y \dot{-} 0) = y = y + (0 \dot{-} y)$.
By $(\dot{-}3)$, $(+4)$ and I.H.

$$suc\,x + (suc\,y \dot{-} suc\,x) = suc(x + (y \dot{-} x))$$
$$= suc(y + (x \dot{-} y)) = suc\,y + (suc\,x \dot{-} suc\,y).$$

Now, 1.1.12 yields $(max1)$.

$(max2)$: $s \dot{-} t = 0$ is by $(max1)$ and $(\dot{-}5)$ equivalent to
$$max\,st = t + (s \dot{-} t) = t + 0 = t.$$

$(\leq trans)$: By the assumptions and $(max2)$, $s = r + (s \dot{-} r)$ and $t = s + (t \dot{-} s)$,
hence by $(+2)$ $t = r + ((s \dot{-} r) + (t \dot{-} s))$ which implies $r \leq t$ by $(\dot{-}4)$.

By $(| - |2)$, an arbitrary equation $s = t$ of type o is seen to be equivalent to
a suitable term, viz. $|s - t|$, being equal to 0. Applying further primitive
recursive functions that have been defined above, this fact is transferred to
arbitrary formulae A of $L(T)$ of type o : to any such A, we can assign a
characteristic term which vanishes exactly when A is the case.

1.1.14 *Recursive definition of a characteristic term* $\chi(A)$ *for formulae A of $L(T)$ of type o*

1. $\chi(s = t) = E_o st$
2. $\chi(\bot) = 1$
3. $\chi(A \wedge B) = max\,\chi(A)\chi(B)$
4. $\chi(A \to B) = (1 \dot{-} \chi(A)) \cdot \chi(B)$

1.1.15 *Proposition*

For any type o formula A of $L(T)$, $T \vdash \chi(A) \leq 1$, and

$$T \vdash A \leftrightarrow \chi(A) = 0$$

Proof of the second claim by subformula induction.
1. By $(| - |2)$ in 1.1.13.
2. $T \vdash \bot \leftrightarrow 1 = 0 \leftrightarrow \chi(\bot) = 0$.
3. By I.H., $T \vdash A \wedge B \leftrightarrow \chi(A) = 0 \wedge \chi(B) = 0 \leftrightarrow max\,\chi(A)\chi(B) = 0$.
4. First observe that for arbitrary $s, t : o$

$$T \vdash (s = 0 \to t = 0) \leftrightarrow (1 \dot{-} s) \cdot t = 0$$

which follows by $(Cases)$ on s. Then, by I.H.,

$$T \vdash (A \to B) \leftrightarrow (\chi(A) = 0 \to \chi(B) = 0) \leftrightarrow \chi(A \to B) = 0.$$

By this result, the logic of T_0 is in fact classical logic: For formulae of $L(T_0)$, there are only two truth values, 0 standing for *true* and 1 standing for *false*. Even bounded quantifiers can be defined for formulae of type o, via their characteristic terms: $Max\, ft = 0$ expresses $\forall x < t\, fx = 0$ and $\Pi ft = 0$ expresses $\exists x < t\, fx = 0$. So, given a characteristic term $\chi(A[x])$ of $A[x]$,

$$max\{\chi(A[x])|x < t\} = \chi(\forall x < tA[x]) \quad and \quad \Pi_{x<t}\chi(A[x]) = \chi(\exists x < tA[x])$$

In particular, the theory $T_{\wedge 0}$ described below in 1.1.19, is an extension by definitions of T_0 for which proposition 1.1.15 holds. For equations of higher type, however, characteristic terms do, in general, not exist in T.

Whereas finite sum, finite product, and finite maximum are straight forward iterations of addition, multiplication, and maximum (of two), respectively, the bounded μ-operator as well as the coding and decoding functions deserve individual attention.

The function $\mu f : o \to o$ grows step by step like the identity I_o, i.e. $\mu fx = x$ (cf. ($\mu 5$) below), until x reaches a zero of f at, say, $x = s$. From there on, μf remains constant (cf. ($\mu 6$)), thus fixing the first zero of f (cf. ($\mu 4$)), if there is any. This may be taken as a guideline for the following more formal argument:

1.1.16   Lemma

The bounded μ-operator has the following properties in T:
($\mu 0$) $\mu f0 = 0$
($\mu 1$) $\mu f(suc\, t) = (\mu ft) + sg(f(\mu ft))$
($\mu 2$) $\mu f(suc\, t) = \mu ft \leftrightarrow f(\mu ft) = 0$
($\mu 3$) $\mu ft \leq t$
($\mu 4$) $\mu ft < t \to f(\mu ft) = 0$
($\mu 5$) $\mu ft = t \leftrightarrow \Pi ft \neq 0$
($\mu 6$) $fs = 0 \wedge s \leq t \to \mu fs = \mu ft$

Proof. ($\mu 0$) and ($\mu 1$) are the recursion equations for the bounded μ-operator. ($\mu 2$) is obvious from ($\mu 1$). ($\mu 3$) follows by induction on t, since ($\mu 1$) implies $\mu f(t + 1) \leq (\mu ft) + 1$. ($\mu 4$) too follows by induction; the induction step runs as follows: $\mu f(suc\, x) < suc\, x$ implies $(\mu fx) + sg(f(\mu fx)) \leq x$, hence $\mu fx = x \to (f(\mu fx) = 0 \wedge \mu f(suc\, x) = \mu fx)$ and by I.H. also $\mu fx < x \to f(\mu fx) = 0 \wedge \mu f(suc\, x) = \mu fx$: In either case we obtain $f(\mu f(suc\, x)) = 0$.

($\mu 5$) is proved by induction on t, using ($\mu 3$). ($\mu 6$) follows by induction on $t \doteq s$, since $f(\mu f s) = 0$ in both cases, by hypothesis, if $\mu f s = s$, and by ($\mu 4$), if $\mu f s < s$.

1.1.17 Lemma: Ordered pairs and their decoding

Ordered pairs and their decoding functions are in T connected as follows:
($op1$) $\langle s, t \rangle = \langle 0, s + t \rangle + s$
($op2$) $\langle 0, s + t \rangle \leq \langle s, t \rangle \leq \langle s + t, 0 \rangle < \langle 0, suc(s + t) \rangle$
($op3$) $(s < s' \rightarrow \langle s, t \rangle < \langle s', t \rangle) \wedge (t < t' \rightarrow \langle s, t \rangle < \langle s, t' \rangle)$
($op4$) $\mu z < suc\langle s, t \rangle (\langle s, t \rangle < \langle 0, suc\, z \rangle) = s + t$
($op5$) $j_0 \langle s, t \rangle = s \wedge j_1 \langle s, t \rangle = t$, $\langle\,,\,\rangle$ *is injective,*
($op6$) $\langle j_0 r, j_1 r \rangle = r$, $\langle\,,\,\rangle$ *is surjective.*

Proof. ($op1$) to ($op3$) are obvious from the definition of ordered pair. By ($op2$), $s + t$ is the smallest z such that $\langle s, t \rangle < \langle 0, suc\, z \rangle$, and $s + t < suc\langle s, t \rangle$. Hence ($op4$) and also $j_0 \langle s, t \rangle = \langle s, t \rangle \doteq \langle 0, s + t \rangle = s$ by ($op2$), $j_1 \langle s, t \rangle = (s + t) \doteq s = t$ which is ($op5$). Finally, since $\langle s + t, 0 \rangle + 1 = \langle 0, suc(s + t) \rangle$, $\langle\,,\,\rangle$ is surjective by ($op1$) and ($op2$) so that ($op5$) implies ($op6$).

From definition 1.1.6 onward, we worked essentially in T_0, the type o fragment of T, and we used linear types only. We now ask whether the schema of primitive recursion in higher types can be extended to tuples of arguments. This question has a simple exact formulation under the tuple notation of 0.1.13 and a simple positive answer if we make use of product types:

1.1.18 Lemma: Simultaneous recursion using product types

Let τ be a type tuple of length $n > 1$. Given term tuples $a : o \rightarrow \tau \rightarrow \tau$ and $b : \tau$, there is a term tuple $f : o \rightarrow \tau$ in T with product types that solves the $2n$ recursion equations

$$f0 = b \text{ and } f(suc\, t) = at(ft)$$

The *proof* that we present here for the sake of brevity makes essential use of product types and consists in a translation of type tuples into product types, term tuples into iterated pairing terms, and back: With any type tuple $\tau = \tau_1, \tau_2, ..., \tau_n$, we associate the product type $\tau^\times = (...(\tau_1 \times \tau_2) \times ... \times \tau_n)$; with any term tuple $b = b_1, b_2, .., b_n$ of type tuple τ, we associate the iterated pairing term $b^\times = (....(b_1, b_2), ..., b_n)$ of type $\tau^\times$; and given $\tau^\times$ as above, for any iterated pairing term $c : \tau^\times$, that implies

$$c_1 = \mathsf{P}_0^{n-1}c, \quad c_i = \mathsf{P}_1(\mathsf{P}_0^{n-i}c) \; for \; 1 < i < n, \quad c_n = \mathsf{P}_1 c :$$

the components c_i of c can be recovered from c, we may put

$$tuple(c) := c_1, ..., c_n \; and \; have \; (tuple(c))^\times \equiv c.$$

Given tuples τ, a and b as in the lemma, we pick fresh variables $x : o$, $y : \tau^\times$ and form $tuple(y) : \tau$ so that $ax\,tuple(y)$ is a term tuple of type tuple τ and $a^* := \lambda xy.(ax\,tuple(y))^\times$ is a term of type $o \to \tau^\times \to \tau^\times$. Applying the recursor R_σ indexed by the product type $\sigma = \tau^\times$, we now define

$$f^* := \mathsf{R}_\sigma a^* b^\times : o \to \tau^\times \; and \; fx := tuple(f^*x) : \tau$$

so that f is nothing but the tuple-version of f^*. The recursion equations for f^* are

$$f^*0 = b^\times \; and \; f^*(suc\,t) = a^*t(f^*t)$$

From these two equations we get the $2n$ equations $f0 = tuple(f^*0) = b$ and

$$f(suc\,t) = tuple(f^*(suc\,t)) = tuple(a^*t(f^*t))$$
$$= tuple((at\,tuple(f^*t))^\times) = at(ft).$$

This proof is more a matter of book keeping than of mathematical argument. It is quite a different question whether the statement of the lemma can be proved in T without product types. In section 1.4, it will be shown that this is the case.

We now turn to an extension of Gödel's theory T by a bounded universal quantifier $\forall x < t$. This extension is needed for the functional interpretation of Heyting arithmetic in all finite types, as will be seen in the next section.

1.1.19   *The extension $T_\wedge$ of T*

The language of $T_\wedge$ extends the language of T by a bounded universal quantifier $\forall x < t$. The inductive definition of *formulae* of T is extended by the clause:

If A is a formula, t is a term of type o, and x is a variable of type o that does not occur in t, then $\forall x < t\,A$ is a formula, called *bounded universal formula*.

$T_\wedge$ has the additional axioms:

B1. $\forall x < 0\,A[x]$

B2. $\forall x < suc\,t\,A[x] \to \forall x < t\,A[x]$

B3. $\forall x < suc\,t\,A[x] \to A[t]$

B4. $\forall x < t\,A[x] \wedge A[t] \to \forall x < suc\,t\,A[x]$

There are no new functionals and no new rules in $T_\wedge$. Axioms and rules of

T are, of course, extended to the language of $T_\wedge$.

This completes the definition of $T_\wedge$, again in a version in linear types only and a version in all finite types. The type o fragment of $T_\wedge$ is denoted by $T_{\wedge 0}$.

Axioms $B2$ to $B4$ might have been collected in the schema
$$\forall x < suc\, t\, A[x] \leftrightarrow \forall x < t\, A[x] \wedge A[t].$$
Following the doctrine of 'short' schemata, however, we prefer the axiomatization given above. In Diller-Nahm [15], the index $\wedge$ was chosen for $T_\wedge$, because bounded universal quantification in arithmetic may be thought of as conjunction of finite length which may, however, depend on parameters. - It must be clear that the expression $x < t$ occurring in $\forall x < t$ is not an atomic formula of our language, and that the $<$ occurring in it is not to be confused with the symbol for the less-relation defined in 1.1.6., though there obviously is a close relation between the two symbols.

We collect some information about the bounded universal quantifier.

1.1.20 Lemma: Laws of bounded universal quantification

(1) If x^o does not occur in A and $T_\wedge \vdash A \wedge B \to C$, then
$T_\wedge \vdash A \wedge \forall x < t\, B \to \forall x < t\, C$
(A and/or B may be missing.)
(2) $\forall x < t\, (A \wedge B) \leftrightarrow \forall x < t\, A \wedge \forall x < t\, B$
(3) $\forall x < s + t\, A \to \forall x < s\, A$
(4) $\forall x < t\, A[x] \to s < t \to A[s]$
(5) $\forall x < s + t\, A[x] \to \forall x < t\, A[s + x]$
(6) $\forall x < t\, x < t$

Proof. (1) to (3), (5), and (6) are proved by induction on t. For (1), (2) and (5), all axioms $B1$ to $B4$ are used. The proof of (3) is an iteration of $B2$. (4) follows from (3) by $B3$. We get the induction step for (6) by applying (1) to $x < y \to x < suc\, y$ and $B4$.

1.1.21 Lemma: Contraction of bounds

If x^o does not occur in $A[y, z]$, $T_\wedge$ proves
$$\forall x < \langle s, Max\, fs \rangle A[j_0 x, j_1 x] \to \forall y < s\, \forall z < fy\, A[y, z]$$
Proof. Let $B := A[j_0 x, j_1 x]$. Since $\langle \ , \ \rangle$ is strictly monotone, we get from lemma 1.1.20 (3) and (4)
$$\forall x < \langle suc\, u, suc\, v \rangle B \to \forall x < \langle suc\, u, v \rangle B \wedge A[u, v]$$

By induction on v, this implies

$$\forall x < \langle suc\, u, Max\, f(suc\, u)\rangle B \to \forall z < Max\, f(suc\, u)\, A[u, z]$$

As above, this yields

$$\forall x < \langle suc\, u, Max\, f(suc\, u)\rangle B \to \forall x < \langle u, Max\, fu\rangle B \wedge \forall z < fu\, A[u, z]$$

Induction on u now yields the conclusion.

1.1.22   *Proposition: Generalized induction rule*

Let a and z be tuples of terms and of variables resp. of the same type tuple, and let y^o not occur in B. Then

$$T_\wedge \vdash B[0, z] \quad and \quad T_\wedge \vdash \forall y < s\, B[x, a] \to B[suc\, x, z] \quad imply \quad T_\wedge \vdash B[t, z]$$

Proof. Put $g := \lambda xyz.a$ and $l := \lambda xz.s$. By lemma 1.1.18 on simultaneous recursion, there is a tuple f of terms such that

$$f0tyz = z \quad and \quad f(suc\, r)tyz = g(t \dot- suc\, r)(j_1 y)(frt(j_0 y)z),$$

and there is a term L such that

$$L0tz = 1 \quad and \quad L(suc\, r)tz = \langle Lrtz, max\{l(t \dot- suc\, r)(frtyz) \mid y < Lrtz\}\rangle$$

After substitution, and by lemma 1.1.20 (1), the first premise becomes

(1) $\forall y < Lttz\, B[0, fttyz]$

Since by lemma 1.1.11 and 1.1.13,

$$s < t \to suc(t \dot- suc\, s) = t \dot- s \quad and \quad s \leq t \to t \dot- (t \dot- s) = s,$$

in the definitions of $f(suc\, r)$ and $L(suc\, r)$, we may put $r = t \dot- suc\, x$ and obtain

$$x < t \to f(t \dot- x)tyz = gx(j_1 y)(f(t \dot- suc\, x)t(j_0 y)z)$$

and

$$x < t \to L(t \dot- x)tz = \langle L(t \dot- suc\, x)tz, max\{lx(f(t \dot- suc\, x)tyz) \mid y < L(t \dot- suc\, x)tz\}\rangle.$$

The last two formulae yield by lemma 1.1.20 (2)

$$x < t \to (\forall y < L(t \dot- x)tz)B[x, f(t \dot- x)tyz] \to$$
$$(\forall y < L(t \dot- suc\, x)tz)(\forall y' < lx(f(t \dot- suc\, x)tyz))B[x, gxy'(f(t \dot- suc\, x)tyz)].$$

Substituting $f(t \dot- suc\, x)tyz$ for z in the second premise, we get

$$(\forall y' < lx(f(t \dot- suc\, x)tyz))B[x, gxy'(f(t \dot- suc\, x)tyz)] \to B[suc\, x, f(t \dot- suc\, x)tyz]$$

The last two formulae yield

(2) $(x \leq t \to (\forall y < L(t \dot- x)tz)B[x, f(t \dot- x)tyz]) \to$
$$(suc\, x \leq t \to (\forall y < L(t \dot- suc\, x)tz)B[suc\, x, f(t \dot- suc\, x)tyz]).$$

Since $t \leq t$, we get from (1) and (2) by *(IND)*

$$(\forall y < L0tz)B[t, f0tyz].$$

By our choice of $L0$ and $f0$ and lemma 1.1.20 (4), the proof is completed.

1.2 Heyting arithmetic in all finite types and its ∧-interpretation

By his Dialectica interpretation, Gödel gave constructive meaning to first order Heyting arithmetic HA. We start here with a larger, though not stronger system of Heyting arithmetic in all finite types HA^ω which we consider to be *the natural span* of the theories HA and T.

1.2.1 Description of Heyting arithmetic in all finite types HA^ω as an extension of the theory $T_\wedge$

(1) The symbols of HA^ω extend the symbols of $T_\wedge$ by the (unbounded) quantifiers ∀ (*for all*) and ∃ (*there is*).

(2) Terms and atomic formulae of HA^ω are the terms and atomic formulae of $T_\wedge$.

(3) The class of formulae of HA^ω is closed under conjunction ∧, implication →, bounded universal quantification $\forall x < t$ and under quantification over variables of arbitrary finite type: If A is a formula and x^τ is a variable of type τ, then $\forall x^\tau A$ and $\exists x^\tau A$ are formulae.

(4) Axioms and rules of $T_\wedge$ are extended by the following axioms and rules for quantification:

$Q1.$ $\forall x^\tau A[x^\tau] \to A[b]$ for terms $b : \tau$ substitutible in A

$Q2.$ $A[b] \to \exists x^\tau A[x^\tau]$ for terms $b : \tau$ substitutible in A

$Q3.$ $A \to B \vdash A \to \forall u^\tau B$, if u^τ is not free in A

$Q4.$ $A \to B \vdash \exists u^\tau A \to B$, if u^τ is not free in B.

This completes the description of HA^ω in all finite types as well as in linear types. Its **type o version** HA_0^ω has as its language the terms and the type o formulae of HA^ω and as its axioms and rules

1. all axioms of $T_{\wedge 0}$ and

2. the axioms and rules of HA^ω, restricted to type o language.

Due to the condition on variables in $Q3$ and $Q4$, the deduction theorem now holds for sentences D only. The substitution theorem holds as in the case of $T_\wedge$ for substitutable terms.

1.2.2 Definability of disjunction and bounded universal quantification

In HA^ω, a **disjunction** $\vee$ may be defined by

$$(A \vee B) :\equiv \exists x^o((x = 0 \to A) \wedge (x \neq 0 \to B)),$$

with x^o not occurring in A, B. The standard laws of disjunction are then derivable in HA^ω:

A8. $A \to A \vee B$ *and* $B \to A \vee B$
A9. $A \to C,\ B \to C \vdash A \vee B \to C,$

and we have the following characterization of the law of excluded middle:

(lem) $A \vee \neg A \leftrightarrow \exists y^o(y = 0 \leftrightarrow A)$

with y^o not occurring in A. HA^ω as well HA_0^ω are extensions of Heyting arithmetic HA. By lemma 1.1.20 (1), (4), and (6), we have

$$\forall x < t\, A \leftrightarrow \forall x(x < t \to A)$$

Hence also the bounded universal quantifier $\forall x < t$ could be introduced as a defined connective in HA^ω.

1.2.3 Lemma: Principle of finite choice

A principle of finite choice is derivable in HA^ω:

$$\forall x < t\, \exists y^\tau A[x, y] \to \exists Y^{o \to \tau} \forall x < t\, A[x, Yx]$$

The *proof* is by induction on t. For $t = 0$, the zero functional $0^{o \to \tau}$ will do. If $\exists Y^{o \to \tau} \forall x < z\, A[x, Yx]$ and $\exists y^\tau A[z, y]$, take such a y_0 and put $Y'x = Yx$ for $x < z$ and $Y'z = y_0$. Then $\forall x < suc\, z\, A[x, Y'x]$. Hence the induction step holds.

1.2.4 Motivating the $\wedge$-translation

As a motivation for a functional translation of an implication $A \to B$ in the language of HA^ω, we consider deduction trees as they are known from natural deduction.

(i) Let a deduction Π of an existential formula $\exists y B[y]$ from a purely universal assumption $\forall w A$ with A in $L(T_\wedge)$ be given. Since the assumption does not contain any existential information, the existential quantifier $\exists y$ must be introduced at some place in Π. This, of course, hinges on the fact that HA^ω does not contain any existential axioms or non-logical rules of inference that introduce existential formulae. We look at the lowest $\exists$-introduction in Π with conclusion $\exists y B[y]$. The premise of this inference is a formula $B[c]$ for some term c which may depend on variables free in $\forall w A$ or in $\exists y B[y]$. Cutting off this conclusion (and all formulae occurrences below it) from Π, we obtain a deduction of $B[c]$ from $\forall w A$.

(ii) In a deduction Π of an essentially quantifier free formula B from an assumption $\forall w A[w]$, the assumption may occur in more than one place at the top of a branch in Π. Working downwards along these branches through Π towards B, the unbounded universal quantifier $\forall w$ must be eliminated from the assumption $\forall w A[w]$ in some place, since the conclusion B does not contain unbounded quantifiers. This again hinges on the fact that HA^ω does not contain axioms or non-logical rules of inference that eliminate unbounded universal quantifiers, like, e.g., the axiom of type-extensionality $\forall x\, fx = gx \to f = g$.
The conclusions of such $\forall$-eliminations will be of the form $A[w_i]$, with w_i depending on the branch in question. Since Π is a finite object, we thus obtain a finite set of instances $A[w_i]$, so-called *challenges* on which B depends. However, taking into account that induction inferences (IND) may occur in Π, the number of instances $A[w_i]$ will depend not only on the number of occurrences of the assumption $\forall w A[w]$, but also on parameters that occur in terms t of the conclusions $F[t]$ of induction inferences (IND). So, we may replace the unbounded assumption $\forall w A[w]$ uniformly by a bounded assumption $\forall x < X A[Wx]$ where $X : o$ is a term possibly depending on parameters in Π and W is a function enumerating the instances w_i for which $A[w_i]$ occurs in Π.
As a consequence, in order to formulate an adequate functional translation of HA^ω, we had to incorporate a bounded universal quantifier $\forall x < t$ into the language of the functional theory T, thus extending T to the theory $T_\wedge$, as has been done in the previous section.

Our motivation for the $\wedge$-translation of implications given here differs sharply from Gödel's motivation for his Dialectica translation which we shall present in the next section.

Following Diller-Nahm [15], we now define the $\wedge$-translation as a functional translation of HA^ω into $T_\wedge$:

1.2.5   *Recursive definition of the* $\wedge$*-translation of* HA^ω

To any formula A of $L(HA^\omega)$, $\wedge$ assigns an expression $A^\wedge \equiv \exists v \forall w A_\wedge[v, w]$ with $A_\wedge[v, w]$ a formula of $L(T_\wedge)$ and tuples v, w of variables as in definition 0.1.15. $A^\wedge$ is in fact a formula of $L(HA^\omega)$.

$(T_\wedge)^\wedge$ $A^\wedge \equiv A$ for $A \in L(T_\wedge)$

Let $A^\wedge$ be as above and $B^\wedge \equiv \exists y \forall z B_\wedge[y, z]$; then

$(\wedge)^\wedge$ $(A \wedge B)^\wedge$ $\equiv \exists v y \forall w z (A_\wedge[v, w] \wedge B_\wedge[y, z])$

$(\to)^\wedge$ $(A \to B)^\wedge$ $\equiv \exists X W Y \forall v z (\forall x < X v z\, A_\wedge[v, W x v z] \to B_\wedge[Y v, z])$
                 in case the tuple w is not empty

$(\to)^\wedge_0$ $(A \to B)^\wedge$ $\equiv \exists Y \forall v z (A_\wedge[v] \to B_\wedge[Y v, z])$ for empty w

$(\forall <)^\wedge$ $(\forall x < t\, B[x])^\wedge \equiv \exists Y \forall z (\forall x < t\, B_\wedge[Y x, z])$

$(\forall)^\wedge$ $(\forall u A[u])^\wedge$ $\equiv \exists V \forall u w\, A_\wedge[u, V u, w]$

$(\exists)^\wedge$ $(\exists u A[u])^\wedge$ $\equiv \exists u v \forall w\, A_\wedge[u, v, w]$

Clearly, the $\wedge$-translation is regular.

For disjunction as defined in 1.2.2 above, we get

$(\vee)^\wedge$ $(A \vee B)^\wedge \equiv \exists x v y \forall w z ((x = 0 \to A_\wedge) \wedge (x \neq 0 \to B_\wedge))$

A formula A coincides with its translation $A^\wedge$ (up to renaming of bound variables) if and only if A is of a form $\exists v \forall w B$ where v, w are tuples of variables and B is a formula of $L(T_\wedge)$. Hence the $\wedge$-translation is idempotent.

1.2.6   *Lemma*

If $A^\wedge \equiv B^\wedge$, then $A \to B$ is $\wedge$-interpretable in $T_\wedge$.

Proof. Using the notation of definition 1.2.5, we have $A_\wedge[v, z] \equiv B_\wedge[v, z]$, and by lemma 1.1.20 (4),

$$T_\wedge \vdash \forall x < 1\, A_\wedge[v, z] \to B_\wedge[v, z]$$

Hence, the functionals $X = \lambda v z.1, W = \lambda x v z.z$, and $Y = \lambda v.v$ $\wedge$-interpret $A \to B$.

1.2.7 $\wedge$*-Interpretation theorem for* HA^ω

The theory HA^ω is $\wedge$-interpretable in $T_\wedge$:

$$HA^\omega \overset{\wedge}{\hookrightarrow} T_\wedge$$

If $HA^\omega \vdash A$ and $A^\wedge \equiv \exists v \forall w A_\wedge[v, w]$, then there is a tuple b of terms from $L(T_\wedge)$ (of the same type tuple as v and with variables in $FV(A)$) for which

$$T_\wedge \vdash A_\wedge[b, w]$$

The *proof by induction on deductions* will be carried out here in some detail, because large parts of it will also be used later (in section 1.7, chapter 3). We start with the axioms and rules of the theory $BLTT$, extended to the language of HA^ω. We use the notation of 0.1.3.3.
$A1$ is an instance of 1.2.6.
$A2$. Let v_0, X, W, Y be given such that by I.H. $A_\wedge[v_0, w]$ and

$$\forall x < Xvz\, A_\wedge[v, Wxvz] \to B_\wedge[Yv, z]$$

are derivable in $T_\wedge$. Substituting v_0 for v and Wxv_0z for w, we obtain $\forall x < Xv_0z\, A[v_0, Wxv_0z]$ by lemma 1.1.20 (1) from the first and $\forall x < Xv_0z\, A[v_0, Wxv_0z] \to B_\wedge[Yv_0, z]$ from the second hypothesis, hence $B_\wedge[Yv_0, z]$ by $A2$. So, the tuple Yv_0 $\wedge$-interprets B.
$A3$. Let all variables that occur in B only be replaced by zero functionals. That does not influence the derivability of $A \to C$. By I.H., there are term tuples X_i, W_i, Y_i for $i = 1, 2$ such that

$$\forall x_1 < X_1v_1z_1\, A_\wedge[v_1, W_1x_1v_1z_1] \to B_\wedge[Y_1v_1, z_1]$$

$$\forall x_2 < X_2v_2z_2\, B_\wedge[v_2, W_2x_2v_2z_2] \to C_\wedge[Y_2v_2, z_2]$$

We have to find term tuples X, W, Y such that
$\forall x < Xvz\, A_\wedge[v, Wxvz] \to C_\wedge[Yv, z]$.
We substitute v for v_1, z for z_2, moreover Y_1v for v_2 and $t := W_2x_2(Y_1v)z$ for z_1. Putting $s := X_2(Y_1v)z$, we obtain by lemma 1.1.20 (1)

$$\forall x_2 < s\, \forall x_1 < X_1vt\, A_\wedge[v, W_1x_1vt] \to \forall x_2 < s\, B_\wedge[Y_1v, t]$$

$$\forall x_2 < s\, B_\wedge[Y_1v, t] \to C_\wedge[Y_2(Y_1v), z]$$

Since by lemma 1.1.21 we have for $r := \langle s, max\{X_1vt \mid x_2 < s\}\rangle$

$$\forall x < r\, A_\wedge[v, W_1(j_1x)v(W_2(j_0x)(Y_1v)z)] \to \forall x_2 < s\, \forall x_1 < X_1vt\, A_\wedge[v, W_1x_1vt]$$

we obtain by two applications of $A3$

$$\forall x < r\, A_\wedge[v, W_1(j_1x)v(W_2(j_0x)(Y_1v)z)] \to C_\wedge[Y_2(Y_1v), z]$$

Hence, the tuples X, W, Y are given by
$Xvz = r$, $Wxvz = W_1(j_1x)v(W_2(j_0x)(Y_1v)z)$, and $Yv = Y_2(Y_1v)$.
$A4$. Writing v for v_1, v_2, we have to find X, W_1, W_2, Y such that

$$\forall x < Xvz(A_\wedge[v_1, W_1xvz] \wedge B_\wedge[v_2, W_2xvz]) \to A_\wedge[Yv, z]$$

The functional tuples X, W_1, W_2, Y given by $Xvz = 1$, $W_1xvz = z$, $W_2xvz = 0$, $Yv = v_1$ will do.

*A*5. By I.H., there are term tuples X, W_1, W_2, Y such that (writing v for v_1, v_2)

$$\forall x < Xvz(A_\wedge[v_1, W_1xvz] \wedge B_\wedge[v_2, W_2xvz]) \rightarrow C_\wedge[Yv, z]$$

We have to find term tuples X_1, X_2, W_1, W_2, Y such that

$$\forall x < X_1vzA_\wedge[v_1, W_1xvz] \rightarrow \forall x < X_2vzB_\wedge[v_2, W_2xvz] \rightarrow C_\wedge[Yv, z]$$

The tuples $X_1 = X_2 = X$ and W_1, W_2, Y given by I.H. $\wedge$-interpret the conclusion, as is seen by distributing $\forall x < Xvz$ over the conjunction according to lemma 1.1.20 (2) and applying *A*3 and *A*5.

*A*6. By I.H., there are term tuples X_1, X_2, W_1, W_2, Y such that (writing again v for v_1, v_2)

$$\forall x < X_1vzA_\wedge[v_1, W_1xvz] \rightarrow \forall x < X_2vzB_\wedge[v_2, W_2xvz] \rightarrow C_\wedge[Yv, z]$$

We have to find term tuples X, W_1, W_2, Y such that

$$\forall x < Xvz(A_\wedge[v_1, W_1xvz] \wedge B_\wedge[v_2, W_2xvz]) \rightarrow C_\wedge[Yv, z]$$

We define X by $Xvz = max(X_1vz)(X_2vz)$ and take over W_1, W_2, Y from the I.H. By lemma 1.1.20 (2) and (3), we get

$$\forall x < Xvz(A_\wedge[v_1, W_1xvz] \wedge B_\wedge[v_2, W_2xvz]) \rightarrow$$
$$(\forall x < X_1vzA_\wedge[v_1, W_1xvz] \wedge \forall x < X_2vzB_\wedge[v_2, W_2xvz])$$

from which we infer the conclusion by *A*3 and *A*6.

*A*7. By I.H., there are term tuples X_i, W_i, Y_i for $i = 1, 2$ such that

$$\forall x < X_1vz_1A_\wedge[v, W_1xvz_1] \rightarrow B_\wedge[Y_1v, z_1]$$
$$\forall x < X_2vz_2A_\wedge[v, W_2xvz_2] \rightarrow C_\wedge[Y_2v, z_2]$$

Taking over Y_1, Y_2 from the I.H. and defining X (again with $z \equiv z_1, z_2$) by

$$Xvz = X_1vz_1 + X_2vz_2,$$

we have to find a term tuple W satisfying

$$(1) \quad \begin{aligned} &\forall x < X_1vz_1 \, Wxvz = W_1xvz_1 \quad \text{and} \\ &\forall x < X_2vz_2 \, W(X_1vz_1 + x)vz = W_2xvz_2, \end{aligned}$$

because for such W, the I.H. implies by 1.1.20 (3) and (5), *A*3 and *A*7

$$\forall x < Xvz \, A_\wedge[v, Wxvz] \rightarrow B_\wedge[Y_1v, z_1] \wedge C_\wedge[Y_2v, z_2]$$

as requested. After first defining a term tuple U by simultaneous recursion 1.1.18

$$Uxvz0 = W_1xvz_1 \text{ and } Uxvz(suc\, y) = W_2(x \dot{-} X_1vz_1)vz_2,$$

we define W by

$$W x v z = U x v z (suc\,x - X_1 v z_1)$$

By lemma 1.1.20 (1) and (6), and since $suc(X_1 v z_1 + x) - X_1 v z_1 = sux\,x$, W satisfies (1) and thus completes the $\wedge$-interpretation of $A7$.

$A\perp$. Since $\perp \to A_\wedge[v, w]$ for any term tuple v, we may choose $v = 0^\sigma$ as a tuple of 0-functionals.

Axioms $(K), (S), (R)$, and, in the presence of product types, (P) are translated by themselves and therefore interpreted trivially.

Rule (Eq). By I.H., there are term tuples X', W', Y' such that

$$\forall x < X'v\; A_\wedge[v, W'xv] \to a = b \text{ and } F_\wedge[a, Y', z]$$

with v, z not occurring in a, b. By (Eq), we infer

$$\forall x < X'v\; A_\wedge[v, W'xv] \to F_\wedge[b, Y', z]$$

So, X, W, Y with $Xvz = X'v, Wxvz = W'xv$, and $Yv = Y'$ $\wedge$-interpret the conclusion of (Eq).

We now turn to the additional axioms and rules of the theory T as stated in 1.1.1. The additional axioms are formulae of T and hence interpreted trivially. It remains to interpret the induction rule, extended to the language of HA^ω:

(IND) $A[0], A[x] \to A[suc\,x] \vdash A[t]$.

By I.H., there are term tuples v_0 and X, W, Y such that

$$A_\wedge[0, v_0, w] \text{ and } \forall y < Xxvz A_\wedge[x, v, Wxyvz] \to A_\wedge[suc\,x, Yxv, z]$$

We have to find v_1 with $A_\wedge[t, v_1, z]$.

We define by simultaneous recursion 1.1.18

$$f0 = v_0 \text{ and } f(suc\,x) = Yx(fx),$$

substitute z for w, fx for v, and obtain

$$A_\wedge[0, f0, z] \text{ and } \forall y < Xx(fx)z A_\wedge[x, fx, Wxy(fx)z] \to A_\wedge[suc\,x, f(suc\,x), z].$$

$A_\wedge[x, fx, z]$ is therefore a formula $B[x, z]$ which for $s := Xx(fx)z$ and $a := Wxy(fx)z$ satisfies the premises of generalized induction 1.1.22. Therefore, by this proposition, $A_\wedge[t, ft, z]$ follows, and ft is the required tuple v_1.

We come to the extension $T_\wedge$ of T as defined in 1.1.19. If we take $\forall x(x < t \to A)$ as a definition of $\forall x < t\,A$ in HA^ω, then $B1$ to $B4$ are theorems of HA^ω and as such do not have to be interpreted separately. However, we prefer to keep the bounded universal quantifier as a connective of its own right also in HA^ω, taking account of it in definition 1.2.5 of the $\wedge$-translation. This has the advantage that in some cases the prefix of a translated formula is shorter than it would become otherwise.

$B1$. $\forall x < 0\,A_\wedge[x, Vx, w]$ is itself an axiom $B1$. We may therefore take 0-functionals for V.

$B2$. $\forall y < Xvz\forall x < suct\,A_\wedge[x, v, Wyxvz] \to \forall x < t\,A_\wedge[x, Yxv, z]$ follows from an axiom $B2$, if we set $Xvz = 1, Wyxvz = z$, and $Yxv = v$.

$B3$. $\forall y < Xvz\forall x < suct\,A_\wedge[x, v, Wyxvz] \to A_\wedge[t, Yv, z]$ follows from an axiom $B3$, if we set $Xvz = 1, Wyxvz = z$, and $Yv = v$.

$B4$. $\forall y < Xvz(\forall x < t\,A_\wedge[x, v_1, W_1yxvz] \wedge A_\wedge[t, v_2, W_2yxvz])$
$\qquad \to \forall x < suct\,A_\wedge[x, Yxv, z]$ with v short for v_1, v_2

follows from an axiom $B4$, if we set $Xvz = 1, W_1yxvz = W_2yxvz = z$, and $Yxv = v_1$ for $x < t$, $Yxv = v_2$ for $t \le x$.

It remains to $\wedge$-interpret the quantifier laws $Q1$ to $Q4$ of HA^ω as stated in 1.2.1.

$Q1$. $\forall x < XvzA_\wedge[Uxvz, v(Uxvz), Wxvz] \to A_\wedge[b, Yv, z]$
holds for $Xvz = 1, Uxvz = b, Wxvz = z$, and $Yv = vb$.

$Q2$. $\forall x < XvzA_\wedge[b, v, Wxvz] \to A_\wedge[Uv, Yv, z]$
holds for $Xvz = 1, Wxvz = z, Uv = b$, and $Yv = v$.

$Q3$ and $Q4$. Given are term tuples X, W, Y that may depend on u such that $\forall x < XvzA_\wedge[v, Wxvz] \to B_\wedge[Yv, z]$. These same term tuples, after abstraction from the variable u, i.e. $\lambda u.X, \lambda u.W, \lambda u.Y$ $\wedge$-interpret $\forall u(A \to B)$ and, at the same time, the conclusions of $Q3$ and $Q4$. This completes the proof of the interpretation theorem.

1.2.8   Comment on the computation of bounds

We comment briefly on how the bounds that are introduced by $(\to)^\wedge$ are transformed or handed down along a formal deduction. We have used the following elementary operations to combine the bounds occurring in the premises of an inference: coding of two consecutive bounds into one in interpreting the chain rule $A3$ by taking a kind of direct product; iteration of this process in interpreting (IND), using proposition 1.1.22; taking the maximum of two bounds in interpreting $A6$; and taking the sum of two bounds in interpreting $A7$. If the bounds given by the interpretation of the premises are ≤ 1, then also the bound used for the interpretation of the conclusion, is ≤ 1, with the sole exception of $A7$: It is addition that generates bounds larger than 1. This reflects a property of our logical system: Rule $A7$ is the only rule in the system that incorporates contraction.

If a bound larger than 1 is needed in the interpretation of a theorem of HA^ω, then rule $A7$ has to be used in any derivation of this theorem.

1.2.9 *Corollary on conservation and relative consistency*

HA^ω *is a conservative extension of* $T_\wedge$. *The consistency of* $T_\wedge$ *implies the consistency of* HA^ω.

1.2.10 *Corollary:* $\wedge$-*Interpretability of* $\{A \leftrightarrow A^\wedge\}$

$$HA^\omega + \{A \leftrightarrow A^\wedge\} \overset{\wedge}{\hookrightarrow} T_\wedge,$$

and this theory is $\wedge$-*maximal relative to* $T_\wedge$.

Both corollaries follow from the theorem; the first because of $(T_\wedge)^\wedge$ in definition 1.2.5 and the second by an application of lemma 0.1.19.

Because the $\wedge$-translation is regular and idempotent, $HA^\omega + \{A \leftrightarrow A^\wedge\}$ is an extension of HA^ω that proves exactly all those formulae that are $\wedge$-interpretable in $T_\wedge$. We axiomatize the schema $\{A \leftrightarrow A^\wedge\}$ by introducing principles that formalize the motivation of the $\wedge$-translation of implication as discussed in 1.2.4: a principle of **independence of premise**, a **Markov principle** adapted to our situation, and the **axiom schema of choice**:

1.2.11 *Definition of the theory* $HA^{\omega+}$

$IP_\wedge$ $\quad (\forall w A \to \exists y B[y]) \to \exists y (\forall w A \to B[y])$ $\qquad$ for $A \in L(T_\wedge)$
$M_\wedge$ $\quad (\forall w A[w] \to B) \to \exists X W (\forall x < X\, A[Wx] \to B)$ for $A, B \in L(T_\wedge)$
$AC_{\sigma,\tau}$ $\forall x^\sigma \exists y^\tau A[x, y] \to \exists Y^{\sigma \to \tau} \forall x\, A[x, Yx]$

Here, w, y are to be read as non-empty tuples of variables of arbitrary type tuple. The bound X is a single variable of type o.
AC designates the union of all schemata $AC_{\sigma,\tau}$.

$$HA^{\omega+} :\equiv HA^\omega + IP_\wedge + M_\wedge + AC$$

$IP_\wedge$ is a formalized version of motivation 1.2.4 (i) for extracting a term from a deduction of $\forall w A \to \exists y B[y]$. $M_\wedge$ formalizes motivation 1.2.4 (ii) for extracting a finite set of challenges from a deduction of $\forall w A[w] \to B$. AC is used to rearrange quantifiers to obtain an $\exists \forall$ prefix.

1.2.12 Lemma

In HA^ω is derivable:

1. $A^\wedge \leftrightarrow A$ *for* $A \in L(T_\wedge)$;
2. $(A \wedge B)^\wedge \leftrightarrow A^\wedge \wedge B^\wedge$;
3. $(\forall x < tB)^\wedge \leftrightarrow \forall x < tB^\wedge$;
4. $(\exists uA)^\wedge \leftrightarrow \exists uA^\wedge$;
5. $(\forall uA)^\wedge \to \forall uA^\wedge$;
6. $(A \to B)^\wedge \to A^\wedge \to B^\wedge$; *furthermore*
7. $HA^\omega + AC \vdash \forall uA^\wedge \to (\forall uA)^\wedge$;
8. $HA^{\omega+} \vdash (A^\wedge \to B^\wedge) \to (A \to B)^\wedge$.

Proof. 1., 2., 4., 5., and 7. are obvious. 3. follows by lemma 1.2.3. We look at 6. and 8. and use the notation of 1.2.5.

$(A \to B)^\wedge \to \forall v \exists y \forall z \exists XW (\forall x < X\, A_\wedge[v, Wx] \to B_\wedge[y, z])$

$\exists XW (\forall x < X\, A_\wedge[v, Wx] \to B_\wedge[y, z]) \to (\forall w A_\wedge[v, w] \to B_\wedge[y, z])$

$\forall z (\forall w A_\wedge[v, w] \to B_\wedge[y, z]) \leftrightarrow (\forall w A_\wedge[v, w] \to \forall z B_\wedge[y, z])$

$\exists y (\forall w A_\wedge[v, w] \to \forall z B_\wedge[y, z]) \to (\forall w A_\wedge[v, w] \to B^\wedge)$

$\forall v (\forall w A_\wedge[v, w] \to B^\wedge) \leftrightarrow (A^\wedge \to B^\wedge)$.

This implies 6. The implications missing for 8. follow by AC, $M_\wedge$, and $IP_\wedge$.

1.2.13 Characterization and extended interpretation theorem for the $\wedge$-translation

(1) *The theories $HA^{\omega+}$ and $HA^\omega + \{A \leftrightarrow A^\wedge\}$ are equivalent:*

$$HA^{\omega+} \equiv HA^\omega + \{A \leftrightarrow A^\wedge\}$$

(2) $HA^{\omega+}$ *is $\wedge$-interpretable in $T_\wedge$:*

$$HA^{\omega+} \overset{\wedge}{\hookrightarrow} T_\wedge$$

(3) $HA^{\omega+}$ *is $\wedge$-maximal relative to $T_\wedge$.*

Proof of (1) $\vdash$: The induction steps for proving $A \leftrightarrow A^\wedge$ by subformula induction are listed in lemma 1.2.12.

$\dashv$: Any instance D of one of the schemata AC, $M_\wedge$, and $IP_\wedge$ is a formula $A \to B$ such that $A^\wedge \equiv B^\wedge$. Hence, by lemma 1.2.6, D is $\wedge$-interpretable in $T_\wedge$, and $HA^\omega \vdash D^\wedge$. Therefore

$$HA^\omega + \{A \leftrightarrow A^\wedge\} \vdash D$$

(2) and (3) follow from (1) and corollary 1.2.10.

1.2.14 Corollary

$HA^{\omega+}$ *is a conservative extension of* $T_\wedge$. *The consistency of* $T_\wedge$ *also implies the consistency of* $HA^{\omega+}$.

We take a closer look at our form of Markov's principle $M_\wedge$. Its motivation as discussed in 1.2.4 (ii) has no relation to Markov's original motivation which exploits the finiteness of all initial segments of the sequence of natural numbers. $M_\wedge$ is a strong principle in that it applies to undecidable, though essentially quantifier free formulae A, B. Given decidability, we get a stronger conclusion.

1.2.15 Lemma

In HA^ω *is derivable (for w not free in B)*

$$\forall w(A \vee \neg A) \to \exists X W(\forall x < X \, A[Wx] \to B) \to \exists w(A \to B)$$

Proof. Given $X : o$ and $W : o \to \tau$, we argue by induction on X. Let $F[X]$ be the above formula without the quantifiers $\exists X W$. $\forall x < 0 \, A[Wx] \to B$ implies $\exists w(A \to B)$ by $B1$, 0.1.4 (2), and $Q2$, hence $F[0]$ holds. For the induction step, we have $\neg A[WX] \to \exists w(A[w] \to B)$ and

$$A[WX] \to (\forall x < suc X \, A[Wx] \to B) \to (\forall x < X \, A[Wx] \to B)$$

by $B4$ and $A3$. So, $F[X]$ implies

$$A[WX] \to (\forall x < suc X \, A[Wx] \to B) \to \exists w(A \to B).$$

Since $A[WX] \vee \neg A[WX]$ by the first antecedent, $F[suc X]$ follows, and (IND) yields the lemma.

It is now easily seen that standard formulations of Markov's principle and of independence of premise follow from the additional axioms listed in definition 1.2.11:

1.2.16 Proposition

In $HA^{\omega+}$ *are derivable Markov's principle*
$M \qquad \forall w(A \vee \neg A) \to \neg \forall w A \to \exists w \neg A$
and the principle of independence of premise
$IP \qquad \forall w(A \vee \neg A) \to (\forall w A \to \exists y B) \to \exists y(\forall w A \to B).$

Proof. By (lem) in 1.2.2,

$$HA^\omega + AC \vdash \forall w(A \vee \neg A) \leftrightarrow \exists Y \forall w(Yw = 0 \leftrightarrow A)$$

For $\forall w(A \leftrightarrow Yw = 0)$, IP is a special case of $IP_\wedge$. An instance of $M_\wedge$ is

$$\neg \forall w\, Yw = 0 \to \exists XW \neg \forall x < X\, Y(Wx) = 0.$$

By the above lemma 1.2.15, we get

$$\neg \forall x < X\, Y(Wx) = 0 \to \exists w \neg Yw = 0.$$

This implies M for $A \leftrightarrow Yw = 0$. That proves the proposition.

Though a reductive functional interpretation like $\wedge$ is not set up to show closure of a theory like HA^ω under specific rules, we do get the admissibility of a Markov rule as an immediate corollary of the interpretation theorem 1.2.7:

1.2.17   *Proposition*

HA^ω *is closed under the following Markov rule:*
(Rule-$M_\wedge$) If A, B are in $L(T_\wedge)$ and $\vdash \forall w A \to B$, then

$$\vdash \exists XW(\forall x < X\, A[Wx] \to B)$$

This is immediate from the $\wedge$-interpretation theorem 1.2.7, because $\exists XW(\forall x < X\, A[Wx] \to B)$ is the $\wedge$-translation of $\forall w A \to B$.

1.3   Dialectica interpretation and equality functionals

Gödel devised his Dialectica translation to interpret Heyting arithmetic HA. The Dialectica translation as such, however, transfers immediately to the language of HA^ω.

1.3.1   *Recursive definition of the Dialectica translation*

To any formula A of HA^ω, the *Dialectica translation* D assigns a formula $A^D \equiv \exists v \forall w A_D[v, w]$ of HA^ω with $A_D[v, w]$ a formula of $L(T)$ and tuples v, w of variables as in definition 0.1.15 as follows:

$$L(T)^D \qquad A^D \quad \equiv \quad A \quad \text{for } A \in L(T)$$

Let A^D be as above and $B^D \equiv \exists y \forall z B_D[y, z]$. Then

$$
\begin{array}{llll}
(\wedge)^D & (A \wedge B)^D & \equiv & \exists vy \forall wz (A_D[v, w] \wedge B_D[y, z]) \\
(\to)^D & (A \to B)^D & \equiv & \exists WY \forall vz (A_D[v, Wvz] \to B_D[Yv, z]) \\
(\forall)^D & (\forall u A[u])^D & \equiv & \exists V \forall uw A_D[u, Vu, w] \\
(\exists)^D & (\exists u A[u])^D & \equiv & \exists uv \forall w A_D[u, v, w]
\end{array}
$$

The bounded universal quantifier $\forall x < t$ is considered a defined connective in this section, defined by $\forall x(x < t \rightarrow ...)$ as in definition 1.2.2. The Dialectica translation D, in particular clause $(\rightarrow)^D$ above, results from the $\wedge$-translation by systematically setting $Xvz = 1$ in $(\rightarrow)^{\wedge}$. Like the $\wedge$-translation, D is regular and idempotent; we even have $A^{D\wedge} \equiv A^D$, but in general, we do not have $A^{\wedge D} \equiv A^{\wedge}$.

1.3.2 *Gödel's motivation for* $(\rightarrow)^D$

Gödel motivates $(\rightarrow)^D$ in a 1972 manuscript (cf. [22]) from which we quote in our notation:
"The right-hand side of $(\rightarrow)^D$ is obtained by stepwise transforming the formula $A^D \rightarrow B^D$ according to the rule that propositions of the form $\exists x(...) \rightarrow \exists y(...)$ (or $\forall y(...) \rightarrow \forall x(...)$, respectively), where x, y may be sequences of variables of any type, are replaced by propositions stating that there exist computable functions which assign to each example for the implicans (or counterexample for the implicatum, respectively) an example for the implicatum (or counterexample for the implicans, respectively), taking account of the fact that $(\neg B_D \rightarrow \neg A_D) \leftrightarrow (A_D \rightarrow B_D)$."
The first part of the argument is equivalent to a principle of independence of premise, the second part to a Markov principle. In view of the closing remark which calls for the stability (or decidability) of B_D, the reference is rather to the principles IP and M in 1.2.16 than to $IP_{\wedge}$ and $M_{\wedge}$ in 1.2.11. Gödel's motivation refers to the language of HA, and the Dialectica translation D certainly is a functional interpretation of HA in T. One may ask, however, whether D also gives a functional interpretation of HA^{ω} in T or, for the sake of stability of the interpreting formulae, at least in $T + \{\neg\neg A \rightarrow A\}$. The answer is negative.

1.3.3 *Example, communicated by W. A. Howard*

In intuitionistic logic, the stabilized form of the law of excluded middle
$$\neg\neg(A \vee \neg A)$$
is derivable. Therefore, as a consequence of *(lem)* in 1.2.2,

(1) $HA^{\omega} \vdash \neg\forall y^o \neg(y = 0 \leftrightarrow u^1 = 0^1)$

The formula (1) is Dialectica translated into
$$\exists y \, \neg\neg(y = 0 \leftrightarrow u^1 = 0^1)$$
However, there is no functional $Y : 1 \rightarrow o$ in T for which

(2) $T \vdash \neg\neg(Yu^1 = 0 \leftrightarrow u^1 = 0^1)$

This follows from the fact that the functionals $Y : 1 \to o$ in T are continuous (cf. definition 1.5.27) which will be shown in theorem 1.5.28, and solutions Y of (2) are not continuous (cf. 1.5.29).
On the other hand, the formula (1) is $\wedge$-translated into

$$\exists XY \neg \forall x < X \neg (Yx = 0 \leftrightarrow u^1 = 0^1),$$

and since

(3) $T_\wedge \vdash \neg \forall y < 2 \, \neg(y = 0 \leftrightarrow u^1 = 0^1),$

the functionals $X = 2$ and $Y = \mathsf{I}$ are $\wedge$-interpreting terms for (1).
In spite of this situation, it is still possible to axiomatize the schema $\{A \leftrightarrow A^D\}$ over HA^ω. If we put in analogy to 1.2.11

IP_D $(\forall w A \to \exists y B[y]) \to \exists y(\forall w A \to B[y])$ for $A \in L(T)$
M_D $(\forall w A[w] \to B) \to \exists w(A[w] \to B)$ for $A, B \in L(T)$,

we obtain a result completely analogous to lemma 1.2.12, only substituting the superscript D for $\wedge$. By the same argument as in the proof of theorem 1.2.13, we obtain:

1.3.4   Characterization lemma for the Dialectica translation

(1) *The theories $HA^\omega + IP_D + M_D + AC$ and $HA^\omega + \{A \leftrightarrow A^D\}$ are equivalent:*

$$HA^\omega + IP_D + M_D + AC \equiv HA^\omega + \{A \leftrightarrow A^D\}$$

(2) *Any formula D-interpretable in T is derivable in $HA^\omega + IP_D + M_D + AC$.*

The second part of this lemma, however, is unsatisfactory, because its reverse direction, the interpretation theorem, fails. How does HA^ω have to be changed in order to allow for a Dialectica interpretation?
Whereas IP_D is simply $IP_\wedge$ with the formula A restricted to the language of T, M_D, in contrast to $M_\wedge$, has an intuitive background (cf. 1.2.4) only for decidable A in which case M_D is implied by $M_\wedge$ according to lemma 1.2.16. However, as Howard's example shows, in order to reach an interpretation theorem for the Dialectica translation, besides decidability, we need characteristic terms for quantifier free formulae. There are two ways to establish this goal:
(i) Restrict the theories HA^ω and T to their type o fragments.
(ii) Extend the theories HA^ω and T by adding equality functionals E_τ with their axioms.

1.3.5   *Definition: Equality functionals*

For a theory Th in finite types, let the **intensional** version $I\text{-}Th$ of Th be the extension of Th by **equality functionals** $\mathsf{E}_\tau : \tau \to \tau \to o$ for $\tau \neq o$ together with the axioms

(E) $\mathsf{E}_\tau ab = 0 \leftrightarrow a = b$ and $\mathsf{E}_\tau ab = 1 \leftrightarrow a \neq b$

E_o is already defined in 1.1.6 by $E_o st = sg|s - t|$, and $E_{o\times\ldots\times o}$ may be defined similarly.

In particular, $I\text{-}T_\wedge$ and $I\text{-}HA^\omega$ are the extensions of $T_\wedge$ and HA^ω, respectively, by equality functionals satisfying axioms (E) (cf. Troelstra [46] and [47], Feferman [19]). We treat both cases (i) and (ii) simultaneously. Equality functionals E_τ are not primitive recursive functionals, if $\tau \neq o \times \ldots \times o$, since $Y := \mathsf{E}_1 0^1$ is a functional that D-interprets the formula (1) in Howard's example 1.3.3.

1.3.6   *Lemma: Characteristic terms in $I\text{-}T_\wedge$*

Any formula A of $T_{\wedge 0}$ as well as of $I\text{-}T_\wedge$ has a characteristic term $\chi(A)$.

For $T_{\wedge 0}$, this is proposition 1.1.15. For $I\text{-}T_\wedge$, we amend the definition 1.1.14 of characteristic terms by the starting clause

$$\chi(a = b) = \mathsf{E}_\tau ab \quad \text{for} \quad a, b : \tau \neq o$$

and proceed again as in proposition 1.1.15.

As already discussed in connection with 1.1.14, these characteristic terms may be used to define bounded quantification in T_0 as well as in $I\text{-}T$ by

$$\forall x < t\, A \equiv max\{\chi(A) \mid x < t\} = 0 \text{ and } \exists x < t\, A \equiv \Pi\{\chi(A) \mid x < t\} = 0$$

We therefore have:

1.3.7   *Corollary*

$T_{\wedge 0}$ is an extension by definitions of T_0, and $I\text{-}T_\wedge$ is an extension by definitions of $I\text{-}T$.

More important: The existence of characteristic terms in the interpreting functional theories leads to the equivalence of the functional translations D and $\wedge$ on the interpreted theories:

1.3.8   *Characterization theorem*

On the basis of HA_0^ω as well as of $I\text{-}HA^\omega$, the following schemata are equivalent:

(1) $A \leftrightarrow A^D$

(2) $IP_D + M_D + AC$

(3) $IP + M + AC$

(4) $IP_\wedge + M_\wedge + AC$

(5) $A \leftrightarrow A^\wedge$

Proof. The equivalence of (1) and (2) is in both cases an immediate consequence of the characterization lemma 1.3.4. The equivalence of (4) and (5) follows in the same way from the characterization theorem 1.2.13. (4) implies (3) by proposition 1.2.16 ; (3) implies (2) because of lemma 1.3.6 and the decidability of equations of type o; (2) implies (4) by corollary 1.3.7.

These equivalences allow us to infer Dialectica interpretation theorems as corollaries to the $\wedge$-intepretation theorem:

1.3.9 *Extended Dialectica interpretation theorems*

$HA_0^{\omega+}$ and $I\text{-}HA^{\omega+}$ are *D*-interpretable in T_0 and $I\text{-}T$, respectively:

$$HA_0^{\omega+} \overset{D}{\hookrightarrow} T_0$$

$$I\text{-}HA^{\omega+} \overset{D}{\hookrightarrow} I\text{-}T$$

Proof. Let Th be either of the theories $HA_0^{\omega+}$, $I\text{-}HA^{\omega+}$, and let FT be the corresponding functional theory $T_{\wedge 0}$, $I\text{-}T_\wedge$, respectively. If $Th \vdash A$, then $Th \vdash A^D$ by the characterization theorem 1.3.8. So, by the extended $\wedge$-interpretation theorem 1.2.13 - which still applies after addition of equality functionals E_τ ($\tau \neq o$) with their qf axioms - there are terms b such that $FT \vdash (A^D)_\wedge[b, w]$. Because of $A^{D\wedge} \equiv A^D$, also $(A^D)_\wedge \equiv A_{D\wedge} \equiv A_D$, and in fact $FT \vdash A_D[b, w]$: the $\wedge$-interpreting terms b of A^D are the *D*-interpreting terms of A. Since the interpreting theories FT are extensions by definitions of T_0 and $I\text{-}T$ respectively, the theorem is proved.

As corollaries, we list further versions and special cases of this theorem:

1.3.10 *Dialectica interpretation theorems*

$$HA_0^\omega \overset{D}{\hookrightarrow} T_0$$

$$HA \overset{D}{\hookrightarrow} T_0$$

$$HA_0^\omega + IP_D + M_D + AC \overset{D}{\hookrightarrow} T_0$$

$$HA_0^\omega + IP + M + AC \overset{D}{\hookrightarrow} T_0$$

$$I\text{-}HA^\omega \overset{D}{\hookrightarrow} I\text{-}T$$

$$I\text{-}HA^\omega + IP + M + AC \overset{D}{\hookrightarrow} I\text{-}T$$

It is the functional interpretation of HA that Gödel explicitly discusses in his 1958 Dialectica paper [21]. The proof that he hints at on the basis of the remark quoted in 1.3.2 is also a basis for a proof of

$$HA_0^\omega \overset{D}{\hookrightarrow} T_0$$

Equality functionals are not mentioned in the 1958 paper. The D- and the $\wedge$-translations of any formula in the respective language are equivalent already in both basic versions of HA^ω which are D-interpretable:

1.3.11 *Proposition*

For $Th = HA_0^\omega$ *as well as for* $Th = I\text{-}HA^\omega$, *if* $A \in L(Th)$, *then*

$$Th \vdash A^D \leftrightarrow A^\wedge$$

Proof. By the characterization theorem 1.3.8, the schema $A^D \leftrightarrow A$, restricted to the respective language, is derivable in $HA_0^{\omega+}$ as well as in $I\text{-}HA^{\omega+}$. By the $\wedge$-interpretation theorem 1.2.7, it is therefore $\wedge$-interpretable in T_0 as well as in $I\text{-}T$. Therefore, for A in $L(Th)$, $(A^D \leftrightarrow A)^\wedge$ and its consequence $A^{D\wedge} \leftrightarrow A^\wedge$ which, by the remark at the end of 1.3.1, is $A^D \leftrightarrow A^\wedge$, are derivable in either version of Th.

E-functionals open up a possibility to transform arbitrary equations into formulae of type o without explicitly mentioning them:

1.3.12 *Lemma*

For terms a, b *of type* τ

$$I\text{-}HA^\omega \vdash a = b \leftrightarrow \forall Y^{\tau \to o}\, Ya = Yb$$

Proof. Already $BLTT$ proves $a = b \to Ya = Yb$ as an instance of (Eq). For the reverse direction, by $(Q1)$, $\forall Y\, Ya = Yb \to (\mathsf{E}a)a = (\mathsf{E}a)b$, and by axiom (E), $0 = \mathsf{E}aa = \mathsf{E}ab \to a = b$. By $A3$, the assertion follows.

Although HA^ω does not prove this equivalence, its type o fragment "does not notice" this defect within the type o language:

1.3.13   Lemma

For terms $a, b : \tau$ and formulae $F[u^\tau]$ of type o,

$$HA_0^\omega \vdash \forall Y^{\tau \to o} \, Ya = Yb \to (F[a] \leftrightarrow F[b])$$

Proof. Since $F[u]$ is of type o, u^τ occurs in $F[u]$ only in type o contexts, say in $c[u] = d[u]$ for finitely many c, d. Now, HA_0^ω proves

$$\forall Y \, Ya = Yb \to c[a] = (\lambda u.c[u])a = (\lambda u.c[u])b = c[b]$$

and similarly for $d[u]$ which implies

$$\forall Y \, Ya = Yb \to (c[a] = d[a] \leftrightarrow c[b] = d[b])$$

by standard laws of equality 0.1.8. The lemma now follows by a straight forward subformula induction.

This lemma suggests a translation of the language of HA^ω into its type-o fragment without introducing equality functionals:

1.3.14   Recursive definition of the type o version A^o for formulae A of $L(I\text{-}HA^\omega)$

1. For $s, t : o$, $(s = t)^o \equiv s = t$
2. For $a, b : \tau \neq o$, $(a = b)^o \equiv \forall Y^{\tau \to o} \, Ya = Yb$
3. For the logical connectives (including $\bot$), the type o version is a homomorphism.

This amounts to substituting $\forall Y \, Ya = Yb$ for equations $a = b$ of type $\tau \neq o$ in the formulae of $L(I\text{-}HA^\omega)$.

The type o version translates $L(I\text{-}HA^\omega)$ into $L(I\text{-}HA_0^\omega)$, $L(HA^\omega)$ into $L(HA_0^\omega)$, and is the identity translation on $L(I\text{-}HA_0^\omega)$.

1.3.15   Proposition (Rath [32])

(1) $I\text{-}HA^\omega \vdash A \leftrightarrow A^o$
(2) $HA^\omega \vdash A$ implies $HA_0^\omega \vdash A^o$
(3) HA^ω is a conservative extension of HA_0^ω.

Proof. (1) follows from lemma 1.3.12 by a straight forward subformula induction due to the homomorphic character of the type o version.

(2) is proved by induction on deductions. If $a = b$ is an equation $(K), (S), (R0), (Rsuc), (P)$ of type $\neq o$, it appears in HA_0^ω "under context" as $c[a] = c[b]$ with $c[u]$ a term of type o. This follows from $(a = b)^o \equiv \forall Y \, Ya = Yb$ as above. All other axioms and rules of HA^ω except (Eq) are type o translated into instances of the same axiom or rule. $(Eq)^o$ becomes $A^o \to \forall Y \, Ya = Yb, F^o[a] \vdash A^o \to F^o[b]$ which follows from lemma 1.3.13.

(3) follows from (2), because the type o version is the identity translation on $L(HA_0^\omega)$.

1.4 Simultaneous recursions in linear types

In the proofs of the $\wedge$- and D-interpretation theorems in the preceding sections, product types played a rôle in one place only, viz. in lemma 1.1.18 on simultaneous recursion. In his Dialectica paper, Gödel does not mention product types. It is in fact possible, even in the combinatorial, non-extensional context of the theory T, to get along without product types and to define simultaneous recursion in linear types only (cf. Diller-Schütte [16], Schütte [36]). We start by introducing a restriction functional by recursion and not just by cases. This functional is extensionally equal to the "naive" restriction functional (cf. (3) in 1.4.2 below), but it enjoys combinatorial properties (cf. (4) in 1.4.2 below) provable in T which, for the "naive" definition of restriction, cannot be proved without extensionality. This restriction functional allows the definition of a course-of-values recursion from which simultaneous course-of-values recursion is derived. This may be considered a detour, but we feel that the transition from course-of-values recursion to its simultaneous version is more transparent than the transition from primitive recursion to simultaneous recursion directly.

1.4.1 *Definition of a restriction functional* $|_\tau$ *of type* $(o \to \tau) \to o \to o \to \tau$

$$|_\tau := \lambda w^{o \to \tau}.R_{o \to \tau}(\lambda x^o u^{o \to \tau} y^o.D_\tau(uy)(wx)(x \dot{-} y))0^{o \to \tau}$$

where D_τ is the case distinction functional introduced in 1.1.5.

We usually write $c|t$ for $|_\tau ct$ which we read c **restricted to (values less than)** t or the **restriction** of c to t.

1.4.2 Lemma: Laws concerning restriction

The restriction functional $|$ *satisfies provably in T:*

(1) $c|0 = 0^{o\to\tau}$

(2) $(c|suc\,t)y = D_\tau((c|t)y)(ct)(t \dot- y),$ *in particular*

(2.0) $t \leq s \to (c|suc\,t)s = ct$

(2.1) $s < t \to (c|suc\,t)s = (c|t)s$

(3) $s < t \to (c|t)s = cs$

(4) $s \leq t \to (c|t)|s = c|s$

Proof. Put for the moment $a := \lambda x u y. D(uy)(cx)(x \dot- y)$, i.e. $|_\tau c = \mathsf{R}a0^{o\to\tau}$.

(1) $c|0 = \mathsf{R}a0^{o\to\tau}0 = 0^{o\to\tau}$

(2) $(c|suc\,t)y = \mathsf{R}a0^{o\to\tau}(suc\,t)y = D((c|t)y)(ct)(t \dot- y)$

Since $t \leq s \to t \dot- s = 0$ and $s < t \to t \dot- s \neq 0$, (2.0) and (2.1) follow.

(3) We show $(c|suc\,s + x)s = cs$ by induction on x:

$(c|suc\,s + 0)s = c(suc\,s)s = cs$ by (2.0), and

$(c|suc\,s + x)s = cs \to (c|suc\,s + suc\,x)s = (c|suc\,s + x)s = cs$ by (2.1).

Since $s < t \to t = suc\,s + (t \dot- suc\,s)$, (3) follows by (IND).

(4) $0 \leq t \to (c|t)|0 = 0^{o\to\tau} = c|0$, and since $suc\,x \leq t \to x \leq t \wedge x < t$,

$$(x \leq t \to \quad (c|t)|x \quad = c|x) \to$$
$$suc\,x \leq t \to (c|t)|suc\,x = \lambda y. D(((c|t)|x)y)((c|t)x)(x \dot- y)$$
$$= \lambda y. D((c|x)y)(cx)(x \dot- y) = c|suc\,x$$

by (3) and (ξ'). (4) now follows by (IND).

In this last step, it is essential for the applicability of (ξ') that the variable y does not occur in subterms replaced by equal subterms under the abstraction λy.

1.4.3 Theorem: Course-of-values recursion

For any term $g : o \to (o \to \tau) \to \tau$, *there is a term* $f : o \to \tau$ *such that*

(CV) $ft = gt(f|t)$

Proof. Given g, we put in analogy to definition 1.4.1:

$$h := \mathsf{R}_{o\to\tau}(\lambda x^o w^{o\to\tau} y^o. D_\tau(wy)(gxw)(x \dot- y))0^{o\to\tau}$$

and

$$fx = h(suc\,x)x$$

We then have $h0 = 0^{o\to\tau}$ and

$$h(suc\,t) = \lambda y. D(hty)(gt(ht))(t \dot- y)$$

Since $t \overset{.}{-} t = 0$, we get

$$ft = h(suc\,t)t = gt(ht)$$

We show by induction on t:

(1) $f|t = ht$.

By 1.4.2 (1), we have $f|0 = 0^{o \to \tau} = h0$ and

$$f|x = hx \to f|suc\,x = \lambda y.D((f|x)y)(fx)(x \overset{.}{-} y)$$
$$= \lambda y.D(hxy)(gx(hx))(x \overset{.}{-} y) = h(suc\,x)$$

By (IND), (1) follows, and the theorem is proved.

1.4.4   Theorem: Simultaneous course-of-values recursion

Given terms $g_i : o \to (o \to \tau_1) \to \ldots \to (o \to \tau_n) \to \tau_i$ $(i = 1, ..., n)$, *there are terms* $f_i : o \to \tau_i$ $(i = 1, ..., n)$ *satisfying in* T *the* n *equations*

$(SimCV)$ $f_i t = g_i t(f_1|t)...(f_n|t)$ $(i = 1, ..., n)$

Proof by induction on n. The case $n = 1$ is theorem 1.4.3.
Let $n > 1$. We use tuple notation and write τ for the type tuple $\tau_1, ..., \tau_{n-1}$ and $w : o \to \tau$ for a tuple of variables $w_1 : o \to \tau_1, ..., w_{n-1} : o \to \tau_{n-1}$. By theorem 1.4.3, there is a term h which satisfies

(2) $hwt = g_n t(w|t)((hw)|t)$

Here, the expression $w|t$ stands abbreviating for the term tuple $w_1|t, ..., w_{n-1}|t$. By I.H., there is a term tuple $f \equiv f_1, .., f_{n-1}$ with

$$f_i t = (\lambda x w.g_i x w((hw)|x))t(f|t) \quad (i = 1, ..., n-1), \text{ i.e.}$$

(3) $f_i t = g_i t(f|t)((h(f|t))|t)$

We put

$$f_n := hf = hf_1...f_{n-1}$$

Then (2) implies

(4) $f_n t = hft = g_n t(f|t)(f_n|t)$

which is the last equation of $(SimCV)$. It remains to show

$$(h(f|t))|t = (hf)|t = f_n|t,$$

because then (3) goes over into the first $n-1$ equations of $(SimCV)$, and the theorem is proved. We show slightly more, i.e. we show by induction on t:

(5) $t \le s \to (h(f|s))|t = f_n|t$

The case $t = 0$ follows by 1.4.2 (1). By 1.4.2 (4), (2) yields

$$t \le s \to h(f|s)t = g_n t((f|s)|t)((h(f|s))|t) = g_n t(f|t)((h(f|s))|t)$$

So, by equation (4),

$$t \leq s \to (h(f|s))|t = f_n|t \to h(f|s)t = f_n t$$

Therefore, $(x \leq s \to (h(f|s))|x = f_n|x) \land x < s$ implies

$$(h(f|s))|(suc\,x) = \lambda y.D(((h(f|s))|x)y)(h(f|s)x)(x \dot- y)$$
$$= \lambda y.D(f_n|x)(f_n x)(x \dot- y) = f_n|(suc\,x), \text{ i.e.}$$

$$(x \leq s \to (h(f|s))|x = f_n|x) \to (x \leq suc\,s \to (h(f|s))|(suc\,x) = f_n|(suc\,x))$$

By (IND), (5) follows, and since $t \leq t$, also the theorem is proved.
As a corollary, we obtain:

1.4.5　Theorem: Simultaneous recursion

Given terms $a_i : o \to \tau_1 \to ... \to \tau_n \to \tau_i$ and $b_i : \tau_i$ $(i = 1, ..., n)$, there are terms $f_i : o \to \tau_i$ $(i = 1, ..., n)$ satisfying the $2n$ equations:

$$
\begin{aligned}
f_i 0 &= b_i \\
(SimR) & \qquad\qquad (i = 1, ...n) \\
f_i(suc\,t) &= a_i t(f_1 t)...(f_n t)
\end{aligned}
$$

Proof. Let τ be the type tuple $\tau_1, ..., \tau_n$ and $w : o \to \tau$ a tuple of variables $w_1, ..., w_n$ and put for $i = 1, ..., n$

$$g_i = \lambda x w.(D(a_i(prd\,x)(w(prd\,x)))b_i x)$$

By the previous theorem, there is a term tuple $f = f_1, .., f_n$ satisfyfing (SimCV), i.e.

$$f_i t = g_i t(f|t) = D(a_i(prd\,t)((f|t)(prd\,t)))b_i t \quad \text{ for } i = 1, ..., n$$

That implies $f_i 0 = b_i$ and, since $prd(suc\,t) = t$ by lemma 1.1.7, as a consequence of 1.4.2 (3),

$$f_i(suc\,t) = a_i t((f|suc\,t)t) = a_i t(ft)$$

By this theorem, we have recovered simultaneous recursion, already proved in lemma 1.1.18 by use of product types, in linear types only. Since lemma 1.1.18 was the only place where product types were used in the $\land$-interpretation of HA^ω (formulated in linear types), we can formulate theorem 1.2.7 also for theories in linear types only:

1.4.6 $\wedge$-*Interpretation theorem for* HA^ω *in linear types*

The theory HA^ω *in linear types is* $\wedge$-*interpretable in* $T_\wedge$ *in linear types:*

$$HA^\omega \overset{\wedge}{\hookrightarrow} T_\wedge$$

where both theories are based on linear types only.

All $\wedge$- and Dialectica interpretation theorems in sections 2 and 3 also hold, if formulated in linear types only.

Theorem 1.4.5 on simultaneous recursion is in fact the key result for an imbedding of the theory T in all finite types into the theory T in linear types only, called *linearization* or *Currying*. For this construction, the conventions on tuple notation introduced in 0.1.13 are quite useful.

1.4.7 *Recursive definition of the linearization of finite types and of terms in these types*

With any finite type τ, we associate a tuple of types $l\tau$, the **linearization of** τ, recursively as follows:

1. $lo \equiv o$
2. $l(\sigma \to \tau) \equiv (l\sigma \to l\tau)$
3. $l(\sigma \times \tau) \equiv l\sigma, l\tau$

For any type τ, $l\tau$ is a tuple of linear types, and if τ is linear, then $l\tau \equiv \tau$. With the conventions 0.1.13 in mind, it is easily seen that, under 2., the length of the tuple $l(\sigma \to \tau)$ is equal to the length of $l\tau$, whereas, under 3., the length of the tuple $l(\sigma \times \tau)$ is equal to the sum of the lengths of $l\sigma$ and $l\tau$. Moreover, we have

$$l((\rho \times \sigma) \to \tau) \equiv l(\rho \to \sigma \to \tau) \quad \text{and}$$

$$l(\rho \to (\sigma \times \tau)) \equiv l(\rho \to \sigma), l(\rho \to \tau)$$

Furthermore, with each variable x^τ of finite type τ, we associate a tuple $x^l \equiv x^{l\tau}$ of distinct variables of type tuple $l\tau$. If τ is linear, we put $x^{l\tau} \equiv x^\tau$. For constants, let

1. $0^l \equiv 0$
2. $suc^l \equiv suc$
3. $\mathsf{K}^l_{\sigma\tau} \equiv \lambda x^{l\tau} y^{l\sigma}.x^l$
4. $\mathsf{S}^l_{\rho\sigma\tau} \equiv \lambda x^{l(\rho \to \sigma \to \tau)} y^{l(\rho \to \sigma)} z^{l\rho}.x^l z^l (y^l z^l)$
5. $(\quad, \quad)^l_{\sigma,\tau} \equiv \lambda x^{l\sigma} y^{l\tau}.x^l, y^l$

where the tuple x^l, y^l is simply the concatenation of the tuples x^l and y^l.

6. $\mathsf{P}^l_{0\sigma\tau} \equiv \lambda x^l y^l . x^l$ and $\mathsf{P}^l_{1\sigma\tau} \equiv \lambda x^l y^l . y^l$

7. For τ a finite type, let x, y be two tuples of variables of the linear type tuples $o \to l\tau \to l\tau$ and $l\tau$, respectively. By theorem 1.4.5 on simultaneous recursion, there is a term tuple f of type tuple $o \to l\tau$ with

$$ f0 = y \text{ and } f(suc\, t) = xt(ft) $$

Then put $\mathsf{R}^l_\tau \equiv \lambda xy.f$.

8. Finally, for $a : \sigma \to \tau$ and $b : \sigma$, $(ab)^l \equiv (a^l b^l)$.

1.4.8   *Lemma*

Let τ be a finite type.

(1) *For a a term of type τ , a^l is a tuple of terms of type tuple $l\tau$.*

(2) *If a is a term of linear type τ, then a^l is again a term of type τ.*

(3) *If a is a term of $L(T)$ in linear types, then $a^l \equiv a$.*

Proof of (1) by subterm induction. If a is a variable or a constant of type τ, a^l is defined outright as a tuple of variables or functionals of type tuple $l\tau$. This is in particular also the case for the pairing and projection functionals (,), P^l_0 and P^l_1. And if $a : \sigma \to \tau$ and $b : \sigma$, then by I.H. $a^l : l\sigma \to l\tau$ and $b^l : l\sigma$ so that $(ab)^l \equiv (a^l b^l) : l\tau$.

(2) is a special case of (1), since $l\tau \equiv \tau$ for linear types τ.

Proof of (3) by subterm induction, with the proviso that all subterms have linear types. $x^{l\tau} \equiv x^\tau$ is correct by definition 1.4.7. For 0, suc, the statement is obvious; for K, S, it follows by $(K'), (S')$ in lemma 0.1.11, similarly for R, since we may choose $f = \mathsf{R}xy$ in 7. of 1.4.7 and obtain $\mathsf{R}^l \equiv \lambda xy.\mathsf{R}xy \equiv \mathsf{R}$ according to definition 0.1.10.

1.4.9   *Proposition*

The theory T in all finite types is a conservative extension of the theory T in linear types only.

Proof. We extend the operation of linearization from types and terms to formulae: For A a formula of the language of $T^\times$, we obtain A^l by replacing in A every equation $a = b$ by the conjunction $(a = b)^l$ of the tuple of equations $a^l = b^l$. By an easy subformula induction, we have $F[a]^l \equiv F^l[a^l]$. For A a formula in linear types only, we have $A^l \equiv A$ by lemma 1.4.8. It therefore suffices to show:

$(*)$ *If T in all finite types proves A, then the theory T in linear types only proves A^l*

Proof by induction on deductions. Since the propositional structure of formulae is not changed by linearization, the induction steps for $A1$ to $A\bot$ go through. Moreover, for any term $t : o$, its linearization t^l is again a (single) term of type o; therefore, the linearization of an instance of (IND) is again an instance of (IND).

(Eq) Let $(a = b)^l$ be the tuple of equations $a_1^l = b_1^l, ..., a_n^l = b_n^l$. Then $A \to (a = b)^l$ is equivalent to (the conjunction of) the n-tuple of formulae $A \to a_i^l = b_i^l$ $(i = 1, ..., n)$. Out of these and $F[a]^l \equiv F^l[a^l]$, we obtain $A^l \to F^l[b^l] \equiv (A \to F[b])^l$ by n applications of (Eq).

$(Ax\ suc)$ $(\neg suc\, t = 0)^l \equiv \neg suc\, t^l = 0$ is again an instance of $(Ax\ suc)$.

(K) $(\mathsf{K}_{\sigma\tau} ab)^l \equiv (\lambda x^\tau y^\sigma . x^\sigma) a^l b^l = a^l$ by (β)

(S) $(\mathsf{S}_{\rho\sigma\tau} abc)^l \equiv (\lambda x^{l\rho \to l\sigma \to l\tau} y^{l\rho \to l\sigma} z^{l\rho} . xz(yz)) a^l b^l c^l = (ac(bc))^l$

(P) $(\mathsf{P}_{0\sigma\tau}(a, b))^l \equiv (\lambda x^{lo} y^{l\tau} . x^{l\sigma})((\lambda x^{lo} y^{l\tau} . x^{lo}, y^{l\tau}) a^l b^l) = a^l$

and analogously for P_1.

$(Ax\ R)$ By our choice of $\mathsf{R}_{l\tau} \equiv \lambda xy.f$ in 1.4.7,7, with f a term tuple solving $(SimR)$, we have $(\mathsf{R} ab0)^l = f[a^l, b^l]0 = b^l$ and

$$(\mathsf{R} ab(suc\, t))^l = f[a^l, b^l](suc\, t^l) = a^l t^l (f[a^l, b^l] t^l) = (at(\mathsf{R} abt))^l$$

Thus, $(*)$ is proved, linearization is shown to be an embedding of T in all finite types into T in linear types only, and proposition 1.4.9 follows. This result easily extends to larger theories:

1.4.10 Corollary

The theory $T_\wedge$ in all finite types is a conservative extension of the theory $T_\wedge$ in linear types only.

Proof. We only have to extend linearization to the language with bounded universal quantifiers and to axioms $B1$ to $B4$. Again, since for any term $t : o$, t^l is a term of type o, the linearizations of axioms $B1$ to $B4$ are instances of $B1$ to $B4$, and the corollary follows.

1.4.11 Corollary

The theory HA^ω in all finite types is a conservative extension of the theory HA^ω in linear types only.

Proof. Unbounded quantifiers $Q \in \{\forall, \exists\}$ over variables of arbitrary finite type are linearized by quantifier tuples, $(Qx^\tau)^l$ is the tuple $Qx^{l\tau}$ of quantifiers. Instances of $Q1$ to $Q4$ are then linearized by tuples of instances of $Q1$ to $Q4$, combined with applications of $A3$, if necessary, and the corollary follows.

Putting the $\wedge$-interpretation theorem 1.2.7 and corollary 1.4.10 together, we obtain:

1.4.12 *Corollary*

HA^ω *in all finite types is a conservative extension of* $T_\wedge$ *in linear types only.*

It is not known whether this result extends to T instead of $T_\wedge$.

1.5 Computability, consistency, continuity

After HA^ω has been shown to be a conservative extension of $T_\wedge$ by finitary means, in order to constructively show the consistency of HA^ω, it suffices to prove the consistency of $T_\wedge$ by constructive means. We quote from Gödel [21], using our notation for types:

"... for the consistency proof of number theory, we can use ... the notion of computable function of finite type on the natural numbers and certain rather elementary principles of construction of such functions. Here, the notion 'computable function of type τ' is defined as follows:
(1) the computable functions of type o are the natural numbers;
(2) if the notions of 'computable function of type τ_0', 'computable function of type τ_1', ... , 'computable function of type τ_k' (with $k \geq 1$) have already been defined, then a computable function of type $\tau_1 \to ... \to \tau_k \to \tau_0$ is defined as an operation, always performable (and constructively recognizable as such), that to every k-tuple of computable functions of types $\tau_1, ..., \tau_k$ assigns a computable function of type τ_0. This notion is to be regarded as immediately intelligible, provided the notions 'computable function of type τ_i' $(i = 0, 1, ..., k)$ are already understood. If we then regard the type τ as a variable, we arrive at the notion, required for the consistency proof, of a computable function of finite type τ."

Tait [43] transforms this notion of computable function of finite type τ into computability predicates for terms of type τ within the theory T. Axioms $(K), (S)$, etc. defining the combinators of T are used to start a *reduction process* by converting the left side of these equations into their right side. We simultaneously introduce the *inside-out* or *strict* version of this process.

1.5.1 *Definition of conversion, redex, reduct, and normal form in the theory T*

(1) The **conversion** relation *conv* is given by

(K)	$\mathsf{K}ab$	*conv*	a
(S)	$\mathsf{S}abc$	*conv*	$ac(bc)$
$(P0)$	$\mathsf{P}_0(a,b)$	*conv*	a
$(P1)$	$\mathsf{P}_1(a,b)$	*conv*	b
$(R0)$	$\mathsf{R}ab0$	*conv*	b
$(Rsuc)$	$\mathsf{R}ab(suc\,t)$	*conv*	$at(\mathsf{R}abt)$

(2) The left sides of the conversions are **redexes**, the corresponding right sides are their **reducts**.

(3) A term is **normal** or **in normal form**, if it does not contain a redex; a redex is **minimal**, if its proper subterms are normal, i.e. if it does not contain another redex.

1.5.2 *Recursive definition of contraction* $a \rightarrow_1 b$, a *contracts to* b

For terms a, b, c, d of suitable type,

(1)   If a *conv* b , then $a \rightarrow_1 b$
(2)   If $a \rightarrow_1 b$, then $ac \rightarrow_1 bc$ and $da \rightarrow_1 db$
(3)   If $a \rightarrow_1 b$, then $(a,c) \rightarrow_1 (b,c)$ and $(d,a) \rightarrow_1 (d,b)$

In the present context, this says that a **contracts to** b , iff b is the result of converting one occurrence of a redex in a; a **strictly contracts** to b, $a \xrightarrow{str}_1 b$, if this redex is minimal.

1.5.3 *Definition of reduction, reduction tree, normalization, strict and strong normalization*

(1) A **reduction chain** is a (finite or infinite) sequence $a_0, ..., a_i, ...$ such that a_i contracts to a_{i+1} for all i in question. a **reduces to** b, $a \rightarrow b$, if there is a reduction chain beginning with a and ending in b; a **strictly reduces to** b, $a \xrightarrow{str} b$, if this reduction chain is **strict**, i.e. if it consists exclusively of strict contractions.

(2) The set of terms b with $a \rightarrow b$, partially ordered by $\rightarrow$, is called the **reduction tree** of a.

(3) A term a **reduces to normal form** b, if b is in normal form and $a \rightarrow b$.

Such a term b is called a **normal form** of a and is often denoted by a^N. a **normalizes** or **reduces to normal form**, if a has a normal form; a **strictly normalizes**, if a strictly reduces to normal form. $a : \tau$ **strongly normalizes**, $a \in SN_\tau$, if every reduction chain starting with a terminates. Clearly, these relations satisfy:

1.5.4   *Lemma*

(1) *Reduction $\rightarrow$ is the reflexive and transitive closure of contraction $\rightarrow_1$.*
(2) *The reduction tree of any term a is a finitely branching tree order.*
(3) *If a term a normalizes, then there is a finite reduction chain from a to a normal form of a.*
(4) *If a term a strongly normalizes, then the reduction tree of a is well founded, hence finite.*
(5) *A term a strongly normalizes, iff all terms b to which a contracts strongly normalize.*
(6) *If $a \rightarrow b$, then $FV(b) \subseteq FV(a)$ and $T \vdash a = b$.*

The last result can be strengthened as only a small fragment of T is needed to prove $a = b$, if a reduces to b. This fragment consists of the axioms for the constants, the equality rule (Eq) and the standard laws for equality collected in lemma 0.1.8. The induction rule (IND) is certainly not a rule of this equational theory.

1.5.5   *Lemma*

(1) *The only functionals of type o in normal form are the numerals.*
(2) *The only functionals of type $\sigma \times \tau$ in normal form are the pairs (a, b) with functionals $a : \sigma$ and $b : \tau$ in normal form.*
(3) *For variables x, a term $xb_1...b_n$ normalizes, iff the terms $b_1, ..., b_n$ normalize. In that case, $(xb_1...b_n)^N \equiv xb_1^N...b_n^N$*
Proof of (1) and (2) by subterm induction.
1. $0 : o$ is normal and a numeral.
2. If $\text{suc}\, t : o$ is normal, then $t : o$ is normal and a numeral by I.H.; so $\text{suc}\, t$ is a numeral.
3. If a normal functional F starts with a constant $\mathsf{K}, \mathsf{S}, \mathsf{R}, \mathsf{P}_i$ or (,), then F, in order to be of type o or of a type $\sigma \times \tau$ has to start with $\mathsf{K}ab$, $\mathsf{S}abc$, $\mathsf{R}abt$, $\mathsf{P}_i c$, or (a, b) respectively. In the first four cases, F starts with a redex - in the third and fourth case t is a numeral and c is an ordered pair by I.H. - and is hence not normal. In the last case, $F \equiv (a, b)$ is of the required

form iff a, b are normal functionals.

(3) Redexes do not start with variables; as subterms of $xb_1...b_n$, they must be subterms of some b_i. By induction along the reduction chains of $xb_1...b_n$ it follows that all reducts of $xb_1...b_n$ are of the form $xb_1'...b_n'$. That proves (3).

For the consistency proof of the theory T and its conservative extensions, it suffices to show that every functional reduces to a unique normal form. We first show uniqueness of normal form for arbitrary reductions.

1.5.6 *Parallel contraction*

In general, contractions do not commute: If $a \to_1 b$ and $a \to_1 c$, then there may not be a term d such that $b \to_1 d$ and $c \to_1 d$. Tait [43] therefore considers *parallel contractions* which have the following intuitive background: If X is a set of occurrences of redexes in a, let

$$a \overset{X}{\to} b$$

indicate that b is the result of contracting the redex occurrences $\in X$ in a inside out. If now

$$a \overset{X}{\to} b \text{ and } a \overset{Y}{\to} c$$

then there is a uniquely determined term d such that

$$a \overset{X \cup Y}{\to} d,$$

and this term d can be reached via b as well as via c, essentially

$$b \overset{Y-X}{\to} d \text{ and } c \overset{X-Y}{\to} d.$$

Formally, parallel contractions are generated recursively:

1.5.7 *Recursive definition of parallel contraction* $\geq_1$

(1) $a \geq_1 a$
(2) Let $a \geq_1 a'$, $b \geq_1 b'$, $c \geq_1 c'$, and $t \geq_1 t'$. Then

(K)	$\mathsf{K}ab$	$\geq_1$	a'
(S)	$\mathsf{S}abc$	$\geq_1$	$a'c'(b'c')$
$(P0)$	$\mathsf{P}_0(a,b)$	$\geq_1$	a'
$(P1)$	$\mathsf{P}_1(a,b)$	$\geq_1$	b'
$(R0)$	$\mathsf{R}ab0$	$\geq_1$	b'
$(Rsuc)$	$\mathsf{R}ab(suc\,t)$	$\geq_1$	$a't'(\mathsf{R}a'b't')$

(3) If $a \geq_1 a'$, $b \geq_1 b'$, then $ab \geq_1 a'b'$.

By (1) and (2), all conversions are parallel contractions; by (3), also

$$a \to_1 b \text{ implies } a \geq_1 b,$$

and since all parallel contractions are reductions, we have

1.5.8 *Lemma*

Reduction $\rightarrow$ *is the transitive closure of parallel contraction* $\geq_1$.

In contrast to contraction, parallel contraction does commute, as sketched above:

1.5.9 *Diamond lemma for parallel contraction*

If $a \geq_1 b$ *and* $a \geq_1 c$, *then there is a term* d *such that*

$$b \geq_1 d \text{ and } c \geq_1 d$$

Proof by induction on definition 1.5.7 of $\geq_1$. If one of the premises, say $a \geq_1 b$, is by (1), then $b \equiv a$, and we may choose $d \equiv c$.

If both of the premises are by (2), or both by (3), say $ab \geq_1 a_1 b_1$ and $ab \geq_1 a_2 b_2$, then the I.H., $a_i \geq_1 a'$ and $b_i \geq_1 b'$ for $i = 1, 2$, immediately gives a common parallel contractum $d \equiv a'b'$.

If one premise is by (2), say $Sabc \geq_1 a_1 c_1 (b_1 c_1)$ and the other is by (3), $Sabc \geq_1 Sa_2 b_2 c_2$, then, by I.H. there are a', b', c' with $a_i \geq_1 a'$, $b_i \geq_1 b'$, and $c_i \geq_1 c'$ for $i = 1, 2$, and

$$a_1 c_1 (b_1 c_1) \geq_1 a'c'(b'c') \text{ as well as } Sa_2 b_2 c_2 \geq_1 a'c'(b'c')$$

follow.

It is readily seen that, in this lemma, also $a \geq_1 d$ holds for the term d constructed in the proof.

1.5.10 *Church-Rosser theorem*

If $a \rightarrow b$ *and* $a \rightarrow c$, *then there is a term* d *such that*

$$b \rightarrow d \quad \text{and} \quad c \rightarrow d$$

Proof. We denote by $\geq_n$ the n-fold concatenation of $\geq_1$, i.e. $a \geq_n b$ iff there are terms a_i $(i \leq n)$ with $a \equiv a_0, b \equiv a_n$, and $a_i \geq_1 a_{i+1}$ for $i < n$. Since reduction $\rightarrow$ is the transitive closure of $\geq_1$, our premises amount to: There are $k, l \in \mathbb{N}$ such that $a \geq_k b$ and $a \geq_l c$.

We prove by induction: There is a term d such that $b \geq_l d$ and $c \geq_k d$. If $k = 0$ or $l = 0$, then $b \equiv a$ or $c \equiv a$, and we may choose $d \equiv c$ or $d \equiv b$, respectively. The induction step is from k, l to $k + 1, l + 1$: There are terms b', c' with $a \geq_1 b' \geq_k b$ and $a \geq_1 c' \geq_l c$. By the diamond lemma, there is a term d' such that $b' \geq_1 d'$ and $c' \geq_1 d'$. Now, applying the I.H. three times, there are terms d_1, d_2, d such that

$$b \geq_1 d_1 \text{ and } d' \geq_k d_1, \ d' \geq_l d_2 \text{ and } c \geq_1 d_2$$

and hence

$$b \geq_1 d_1 \geq_l d \text{ and } c \geq_1 d_2 \geq_k d$$

which proves the assertion.

1.5.11 Corollary: Uniqueness of normal form

Every term a has at most one normal form a^N.

Proof. Let b, c be terms in normal form to which a reduces. By the Church-Rosser theorem, b and c reduce to the same term d. However, b and c are normal, hence irreducible. Therefore $b \equiv d \equiv c$ is the unique normal form a^N of a.

1.5.12 Lemma

Let b be a strongly normalizing term tuple of type tuple ρ, and τ a type.
(1) $0^{\rho \to \tau} b$ strongly normalizes to 0^τ.
(2) If $\tau = \tau_1 \times \tau_2$, then 0^τ is the only pair to which $0^{\rho \to \tau} b$ reduces.

Proof by induction on the length of ρ and side induction on the size of the (finite) reduction trees of the terms b.
1. For empty ρ, $0^{\rho \to \tau} \equiv 0^\tau$ which is normal.
2. For $\rho = \rho_0, \rho'$ with ρ_0 a type, $b \equiv b_0, b'$, either
2.1. $0^{\rho \to \tau} b \to_1 0^{\rho' \to \tau} b'$, which is necessarily the case when all terms of b are normal, or
2.2. $0^{\rho \to \tau} b \to_1 0^{\rho \to \tau} c$, where some $b_i \to_1 c_i$, the other $b_j \equiv c_j$ remaining unchanged.
(1) Now follows by I.H. in case 2.1 and by side I.H. in case 2.2.
(2) Follows similarly: For $\tau = \tau_1 \times \tau_2$, $0^{\rho \to \tau} b$ contracts to a pair only in case 2.1 for empty ρ', and that pair is 0^τ, as claimed.

Up to this point, types did not really enter into the discussion of this section. For the definition of computability predicates, however, they are the decisive measure, as is expressed in the above quote from Gödel [21]. We introduce computability predicates relative to strong normalization.

1.5.13   Recursive definition of strong computability predicates $SComp_\tau$ for terms in all finite types τ

(1) $t \in SComp_o$ iff $t : o$ strongly normalizes.

(2) $a \in SComp_{\sigma \to \tau}$ iff for all $b \in SComp_\sigma$, $(ab) \in SComp_\tau$.

(3) $a \in SComp_{\sigma \times \tau}$ iff $a : \sigma \times \tau$ strongly normalizes and, for all $b : \sigma, c : \tau$, if $a \to (b, c)$ then $b \in SComp_\sigma$ and $c \in SComp_\tau$.

A term $a : \tau$ is **strongly computable**, if a belongs to $SComp_\tau$. A term tuple is **strongly computable**, if its components are strongly computable.

1.5.14   Lemma

$SComp_\tau$ *is closed under reduction:*
Let $a, b : \tau$ and $a \to b$. Then $a \in SComp_\tau$ implies $b \in SComp_\tau$.

Proof by induction on τ. 1. By definition, $SComp_o = SN_o$.

2. Let $a \in SComp_{\sigma \to \tau}$ and $a \to b$. Then, for all $c \in SComp_\sigma$,
$ac \in SComp_\tau$ and $ac \to bc$.
So, by I.H., $bc \in SComp_\tau$ for all $c \in SComp_\sigma$, i.e. $b \in SComp_{\sigma \to \tau}$.

3. Let $a \in SComp_{\sigma \times \tau}$ and $a \to b$. Then $a \in SN_{\sigma \times \tau}$ by definition and hence $b \in SN_{\sigma \times \tau}$ by lemma 1.5.4 (5). Moreover:
If $b \to (c, d)$, then also $a \to (c, d)$, and therefore $c \in SComp_\sigma$ and $d \in SComp_\tau$; so $b \in SComp_{\sigma \times \tau}$

By iterating (2) of definition 1.5.13 of strong computability, it is easily seen:

1.5.15   Lemma

For any type tuple ρ and types σ, τ,

(1) $a \in SComp_{\rho \to o}$ iff for all $b \in SComp_\rho$ $(ab) \in SN_o$

(2) $a \in SComp_{\rho \to \sigma \times \tau}$ iff for all $b \in SComp_\rho$ $(ab) \in SN_{\sigma \times \tau}$ and also, if $(ab) \to (c, d)$, then $c \in SComp_\sigma$ and $d \in SComp_\tau$.

This follows by induction on the length of the type tuple ρ.

1.5.16   Proposition: Strong normalization of strongly computable terms

For any type τ,

(1) $a \in SComp_\tau$ implies $a \in SN_\tau$

(2) $0^\tau \in SComp_\tau$

Proof by induction on types. Any type τ has a shape $\rho \to o$ or $\rho \to \tau_0 \times \tau_1$

with ρ a tuple of types. Let b be a term tuple of type tuple ρ and
$b \in SComp_\rho$. By I.H., the terms in b strongly normalize.
1. If $\tau = \rho \to o$, by lemma 1.5.12 (1), $0^\tau b : o$ also strongly normalizes, i.e.
it is strongly computable. Therefore, 0^τ is strongly computable by lemma
1.5.15 (1).
2. If $\tau = \rho \to \tau_0 \times \tau_1$, by lemma 1.5.12 (1) and (2), $0^\tau b : \tau_0 \times \tau_1$ strongly
normalizes, and the only pair to which it reduces is $0^{\tau_0 \times \tau_1} \equiv (0^{\tau_0}, 0^{\tau_1})$ where
both 0^{τ_i}, being normal, are strongly computable by I.H. So, 0^τ is strongly
computable by lemma 1.5.15 (2).
That proves (2). If 0^ρ denotes the tuple of 0-functionals of type tuple ρ,
$0^\rho \in SComp_\rho$. Therefore, for $a \in SComp_\tau$, $a0^\rho$ is strongly computable by
lemma 1.5.15 and hence also strongly normalizing by definition, due to the
maximal choice of ρ. Then, certainly, also a strongly normalizes, which
proves (1).

1.5.17 *Strong computability theorem*

Every term of the theory T is strongly computable.

Proof by subterm induction. Since strong computability is closed under
application, it suffices to prove the theorem for the constants and variables
of T.
1. $0 : o$ is in $SComp_o$.
2. As any term $t \in SComp_o$ strongly normalizes to a term t^N, $suc\, t$ strongly
normalizes to $suc\, t^N \equiv (suc\, t)^N$. So suc is in $SComp_1$.
For the next points, let the type $\tau = \pi \to o$ or $\tau = \pi \to \tau_0 \times \tau_1$ with π a
type tuple, and let d be a strongly computable term tuple of type tuple π.
3. Any term $x^\tau d$ strongly normalizes to $x^\tau d^N$ and does not reduce to a pair.
So, x^τ is strongly computable.
4. For $a \in SComp_\tau$, $b \in SComp_\sigma$, any reduction chain of the term $\mathsf{K}_{\sigma\tau} abd$
contains a contraction

$$\mathsf{K}_{\sigma\tau} a'b'd' \to_1 a'd'$$

where a, b, d reduce to a', b', d', and $a'd'$ strongly normalizes to $(ad)^N$. So
also $\mathsf{K}_{\sigma\tau} abd$ strongly normalizes to $(ad)^N$. Moreover, if $\mathsf{K}_{\sigma\tau} abd$ reduces to
a pair (e, f), then also the strongly computable term ad reduces to (e, f),
and e, f are strongly computable. Therefore, the term $\mathsf{K}_{\sigma\tau} abd$ and also the
combinator $\mathsf{K}_{\sigma\tau}$ are strongly computable.
5. For terms $a \in SComp_{\rho\to\sigma\to\tau}$, $b \in SComp_{\rho\to\sigma}$, $c \in SComp_{\rho\to\tau}$, any
reduction chain of the term $\mathsf{S}_{\rho\sigma\tau} abcd$ contains a contraction

$$\mathsf{S}_{\rho\sigma\tau} a'b'c'd' \to_1 a'c'(b'c')d'$$

where a, b, c, d reduce to a', b', c', d', and $a'c'(b'c')d'$ strongly normalizes to $(ac(bc)d)^N$. So also $\mathsf{S}_{\rho\sigma\tau}abcd$ strongly normalizes to $(ac(bc)d)^N$. Moreover, if $\mathsf{S}_{\rho\sigma\tau}abcd$ reduces to a pair (e, f), then also the strongly computable term $ac(bc)d$ reduces to (e, f), and e, f are strongly computable. Therefore, the term $\mathsf{S}_{\rho\sigma\tau}abcd$ and also the combinator $\mathsf{S}_{\rho\sigma\tau}$ are strongly computable.

6. Applied to terms $a \in SComp_\sigma$, $b \in SComp_\tau$, the pairing functional $(\, , \,) : \sigma \to \tau \to \sigma \times \tau$ yields the pair (a, b) which strongly normalizes to (a^N, b^N). Any pair (a', b') to which (a, b) reduces, has strongly computable components a', b' by 1.5.14. So, (a, b) and also $(\, , \,)$ are strongly computable.

7. Consider the projection functional $\mathsf{P}_1 : \sigma \times \tau \to \tau$ and let $c \in SComp_{\sigma \times \tau}$.

7.1. Either c does not reduce to a pair. In that case, $\mathsf{P}_1 cd$ strongly normalizes to $\mathsf{P}_1 c^N d^N$ which is not a pair.

7.2. Or c reduces to a pair. In that case, every reduction chain of $\mathsf{P}_1 cd$ contains a contraction

$$\mathsf{P}_1(c_0, c_1)d' \to_1 c_1 d'$$

where c, d reduce to $(c_0, c_1), d'$. Then $c_1 \in SComp_\tau$ and $d' \in SComp_\pi$ so that $c_1 d'$ is strongly computable, and $\mathsf{P}_1 cd$ strongly normalizes. Moreover, if $\mathsf{P}_1 cd$ reduces to a pair (e, f), then by 7.1, also c reduces to a pair (c_0, c_1) such that

$$\mathsf{P}_1 cd \to c_1 d \to (e, f)$$

So e, f and therefore also $\mathsf{P}_1 cd$ are strongly computable.

Taking 7.1 and 7.2 together, P_1 is strongly computable. P_0 is treated analogously.

8. For $s : o$, let $suc^0 s \equiv s$ and $suc^{k+1}s \equiv suc\,(suc^k s)$. Any strongly computable, i.e. strongly normalizing term $t : o$ reduces to a normal term $suc^k s$ where $s : o$ does not start with suc.

Now let a, b, t, d be strongly computable. The proof of strong computability of $r :\equiv \mathsf{R}_\tau abtd$ is by induction on k. For $k = 0$, there are two cases analogous to 7 :

8.1. $k = 0$ and $t^N \neq 0$. Then r normalizes strongly to $\mathsf{R}_\tau a^N b^N t^N d^N$ which is not a pair. So r is strongly computable.

8.2. $k = 0$ and $t^N = 0$. In that case, every reduction chain of r contains a contraction

$$\mathsf{R}_\tau a'b'0d' \to_1 b'd'$$

where a, b, d reduce to a', b', d', and r strongly normalizes. Moreover, if r reduces to a pair (e, f), then there are also such b', d' such that $b'd'$ reduces to (e, f) so that e, f are strongly computable. Then also r is strongly computable.

8.3. $k > 0$, i.e. $t^N \equiv suc\, s$. In analogy to 8.2, every reduction chain of r contains the contraction

$$\mathsf{R}_\tau a'b'(suc\, s')d' \to_1 a's'(\mathsf{R}_\tau a'b's')d'$$

where $a \to a', b \to b', t \to suc\, s', d \to d'$, and by I.H. also $r' :\equiv \mathsf{R}_\tau a'b's'$ is strongly computable. So r strongly normalizes. Moreover, if r reduces to a pair (e, f), then also some $a's'r'd'$ will reduce to (e, f) whence the components e, f are strongly computable. So also r is strongly computable. Taking 8.1, 8.2, 8.3 together, it follows by induction on k that R_τ is strongly computable.

The theorem now follows by subterm induction.

If we restrict the assertion of the theorem to linear types - which would suffice for most results in the sequel - the argument would become shorter and more transparent.

This theorem, lemma 1.5.5, and corollary 1.5.11 immediately imply:

1.5.18   *Strong normalization theorem*

(1) *Every term a of the theory T strongly normalizes to a unique normal form a^N, recursively computable from a, and $T \vdash a = a^N$.*
(2) *Every functional of type o strongly normalizes to a numeral.*
(3) *Every functional of type $\sigma \times \tau$ strongly normalizes to a pair of normal functionals.*

1.5.19   *Comment: Constructive vs. finitary*

The arguments leading to these theorems are, though combinatorial, not purely finitary. The relations of conversion, contraction, parallel contraction, and the concept of reduction chain of length n are primitive recursive. Therefore, the proofs of the diamond lemma and the Church-Rosser theorem are finitary arguments, and the normalization procedure, the metamathematical function $a \to a^N$, is a recursive function. The concept $SComp_o$ of normalizability, the fact that a term possesses a reduction chain that leads to its normal form, however, is defined as a recursively enumerable or Σ_1^0 predicate. Thus, $SComp_1$ is already a Π_2^0 predicate, and with growing type τ, the predicates $SComp_\tau$ run up through the entire arithmetical hierarchy. In Gödel's words: "If we then regard the type τ as a variable, we arrive at the notion, required for the consistency proof," of computability of functionals of finite type, and this notion is no longer a predicate in the arithmetical hierarchy: It cannot be expressed by a formula of first order arithmetic. It is this extension of the finitary standpoint that Gödel refers to in the title of his Dialectica paper.

There are other proofs of the normalization theorem which, like Howard [24], use transfinite induction up to the first ϵ-number ϵ_0. [10] contains a proof of strong normalization, making use of transfinite induction up to the first ω-critical number. All such proofs transcend the proof power of first order arithmetic which is unavoidable due to Gödel's second incompleteness theorem, because the normalization theorem implies the consistency of T and $T_\wedge$ and hence - because of the interpretation theorems - of HA^ω and $HA^{\omega+}$ by finitary means.

A consequence of the fact that functionals of type o reduce to numerals is:

1.5.20   *Proposition*

If a sentence A is provable in $T_\wedge$, then A has a closed proof in $T_\wedge$, i.e. a proof consisting of sentences only and not using the induction rule (IND).

Proof by induction on deductions. If A is an axiom, then A, being closed, is a closed proof of itself. If A is a conclusion of an inference $A5, A6, A7$, then the premise of this inference is a sentence, too, and has a closed proof without (IND) by I.H. So, A has such a proof, too. If A is a conclusion of an inference $A2, A3, (Eq)$, the free occurrences of variables x^τ that may occur in the premises of this inference, are substituted by zero-functionals 0^τ of the same type τ. By the substitution theorem, these now closed premises have proofs of the same structure as before. Therefore, by I.H., they have closed proofs without (IND), and so has A.

Let $A \equiv F[t]$ be the conclusion of an inference (IND) with premises $F[0], F[x] \to F[suc\,x]$. If x does not occur in $F[x]$, then $A \equiv F[0]$ has a closed proof by I.H. If x does occur in $F[x]$, t is a functional of type o, and therefore by the normalization theorem

$$T \vdash n = t \text{ for some numeral } n$$

with a closed proof without (IND), as remarked following lemma 1.5.4. For any numeral $i < n$, $F[i] \to F[suc\,i]$ is a sentence with a proof of the same structure as the proof of $F[x] \to F[suc\,x]$. So, by I.H., for any $i < n$, $F[i] \to F[suc\,i]$ has a closed proof without (IND). These proofs are put together by n inferences $A2$

$$F[i], F[i] \rightarrow F[suc\,i] \vdash F[suc\,i]$$

thus rendering a closed proof without (IND) of $F[n]$. By an inference (Eq)

$$n = t, F[n] \vdash F[t]$$

we finally obtain the desired proof of $F[t] \equiv A$.

This immediately implies that, restricted to sentences, the theory T has the same expressive power as the theory $T_\wedge$:

1.5.21 Corollary

Restricted to sentences, $T_\wedge$ is an extension by definitions of T.

Proof. For functionals $t : o$, t^N is a numeral and

$$T_\wedge \vdash \forall x < tA[x] \leftrightarrow \top \quad \text{if } t^N = 0$$

$$T_\wedge \vdash \forall x < tA[x] \leftrightarrow A[0] \wedge \ldots \wedge A[n] \quad \text{if } t^N = suc\,n$$

In sentences B of $L(T_\wedge)$, we may therefore take the left sides of these equivalences as abbreviations of their right sides, starting with the outermost bounded quantifier, and thus obtain a sentence B' of $L(T)$. If $T_\wedge \vdash B$, then B has a closed proof Π in which all sentences of $T_\wedge$ that occur in it may be looked at as abbreviations of sentences of $L(T)$. The proof Π will thus be transformed into a proof Π' of B' in T.

In spite of this consequence of computability, it remains an open problem whether $T_\wedge$ is in general a conservative extension of T.

The normal functionals lend themselves to a construction of the type structure NF of *Functionals in Normal Form* over the set $\mathbb{N}$ of numerals which differs considerably from the full type structure $\mathbb{N}^\omega$ (cf. Troelstra [46] where NF is introduced as $CTNF$):

1.5.22 Definition of the type structure NF of Normal Functionals

$$NF_\tau := \{c : \tau \mid c \text{ is a normal functional of } T\}$$

For any constant $C : \tau$,

$$C_{NF} :\equiv C \in NF_\tau$$

For $a \in NF_{\sigma \rightarrow \tau}$ and $b \in NF_\sigma$

$$(ab)_{NF} := (ab)^N$$

Closely related to this type structure is the Closed Term Model CTM which, however, interprets equality non-trivially and does therefore not conform with definition 0.1.2:

$$CTM_\tau := \{c : \tau \mid c \text{ is a functional of } T\}$$

For any constant $C : \tau$,

$$C_{CTM} :\equiv C \in CTM_\tau$$

For $a \in CTM_{\sigma \to \tau}$ and $b \in CTM_\sigma$

$$(ab)_{CTM} := (ab)$$

For $a, b \in CTM_\tau$,

$$CTM \models a = b \Leftrightarrow a^N \equiv b^N$$

NF may be obtained from CTM by factorizing CTM over this equivalence relation.

1.5.23   Lemma

(1) $NF_o = \mathbb{N}$ *the set of numerals.*
(2) *For all functionals a, $a_{NF} \equiv a^N$.*

Proof. (1) is lemma 1.5.5 (1). (2) is immediate by induction on the length of the functional a, since all constants are normal.

1.5.24   Soundness theorem for NF

NF is a recursive model of $T_\wedge$ in the following sense:
(1) *NF is a model of $T_\wedge$.*
(2) *The sets NF_τ are primitive recursive sets of functionals of $L(T)$; the interpretation of constants of $L(T)$ is identity; the interpretation of application $(\ )_{NF}$ is recursive.*
(3) *For sentences A of $L(T_\wedge)$, it is decidable whether $NF \models A$.*

Proof. (1) By proposition 1.5.20, every sentence provable in $T_\wedge$ has a closed proof that does not use (IND). The argument is by induction on such a proof.
The axioms that give rise to conversions are of the form $a = b$ with $a^N \equiv b^N$ and hence true in NF. Axioms and rules $A1$ to $A\bot$, (Eq) are classically correct and are hence true, respectively correct, in NF. Axioms (suc) and $B1$ to $B4$ hold in NF because of lemma 1.5.23 (1).

(2) is already discussed in comment 1.5.19.

(3) is shown by subformula induction. For functionals $a, b : \tau$,

$$NF \models a = b \Leftrightarrow a^N \equiv a_{NF} \equiv b_{NF} \equiv b^N,$$

and $a^N \equiv b^N$ is decidable, because, given functionals a, b of the same type, a^N and b^N are recursively computable from a, b by normalization and uniquely determined by the Church-Rosser theorem. Conjunction and implication are primitive recursive operations; bounded universal quantification, by the argument in corollary 1.5.21, is a recursive operation. All three of them preserve decidability.

It is this last point that gives constructive content to the following:

1.5.25   *Consistency theorem*

The theories $T, T_\wedge, HA^\omega, HA^{\omega+}$ *as well as their type o fragments are consistent.*

Proof. Since $NF \not\models \bot$, $T_\wedge$ is consistent by the soundness theorem. By corollary 1.2.14 to the extended $\wedge$-interpretation theorem, $HA^{\omega+}$ is a conservative extensions of $T_\wedge$ and therefore consistent, too. Hence, all theories mentioned in the theorem, being subtheories of $HA^{\omega+}$, are consistent.

With this result, the program of Gödel's Dialectica paper is completed. We add, however, some further immediate consequences of the above results.

1.5.26   *Corollaries to normalization and soundness*

Soundness together with lemma 1.5.4 immediately implies:

(1) *For functionals* a, b

$$T_\wedge \vdash a = b \ \Leftrightarrow \ a^N \equiv b^N \ \Leftrightarrow \ NF \models a = b$$

(2) *For functionals* a, b, *it is decidable whether* $T_\wedge \vdash a = b$

For functionals of type o, (1) (and (2)) extend to inequalities, because by lemma 1.1.13 ($| - |2$)

$$T \vdash \neg s = t \leftrightarrow sg|s^N - t^N| = 1 :$$

(3) *For functionals* $s, t : o$

$$T_\wedge \vdash \neg s = t \ \Leftrightarrow \ s^N \not\equiv t^N$$

Equivalence (1), however, does not extend to open equations between terms of type o :

(4) *There are normal terms* s, t *of type* o *such that*

$$T_\wedge \vdash s = t \text{ hence } NF \models s = t \text{ yet } s \not\equiv t$$

An example is $s \equiv 0 + x$, $t \equiv x$. That $0 + x = x$ is *valid* in NF means only that for all numerals n, $(0 + n)^N \equiv n$. The derivation of $0 + x = x$ in T uses the induction rule (IND). Restriction to closed derivations according to proposition 1.5.20 avoids applications of (IND) and thereby also avoids this phenomenon.

If λ-abstraction is applied to these provably equal terms $s \equiv 0 + x$, $t \equiv x$, we obtain different functionals $\lambda x. + 0x \equiv +0$ and $\lambda x. x \equiv \mathsf{I}$ of type 1 in normal form, so that $NF \not\models +0 = \mathsf{I}$ and by (1), $T_\wedge \not\vdash +0 = \mathsf{I}$. However, by rule (ξ) (cf. 0.1.14), $0 + x = x$ implies $\lambda x. + 0x = \lambda x. x$, i.e. $T + (\xi) \vdash +0 = \mathsf{I}$:

(5) T *and* $T_\wedge$ *are not closed under rule* (ξ) *and hence not closed under the rule* $(T - EXT)$ *of type extensionality.*

However, $T_\wedge + (\xi)$ is consistent, because the full type structure $\mathbb{N}^\omega$ is a model even of $HA^\omega + (t - ext)$ (cf. lemma 0.1.14.1 and the remark above it). Therefore:

(6) $T_\wedge$ *is incomplete for closed equations of type 1:*

$$T_\wedge \not\vdash +0 = \mathsf{I} \text{ as well as } T_\wedge \not\vdash +0 \neq \mathsf{I}$$

This also means that (3) does not extend to inequalities of type 1.
The full extensional model also yields a criterion for the derivability of inequalities of type 1:

(7) *For functionals* $f, g : 1$

$$T_\wedge \vdash f \neq g \Leftrightarrow T_\wedge \vdash fk \neq gk \quad \text{for some } k \in \mathbb{N}$$

Proof. The implication from right to left is immediate. Given the left side, $\mathbb{N}^\omega \models f \neq g$. Due to the extensionality of this model, there must be a numeral k such that $\mathbb{N}^\omega \models fk \neq gk$. But since $\mathbb{N}^\omega_o = \mathbb{N}$, this is $(fk)^N \not\equiv (gk)^N$. By (3), this implies $T_\wedge \vdash fk \neq gk$.

Already for primitive recursive functions $f : 1$, given their primitive recursive definition, it is undecidable whether f is extensionally equal to the 0-function. Since all such f are definable in T, (7) implies for $g \equiv 0^1$:

(8) *It is recursively undecidable whether* $T_\wedge \vdash f \neq 0^1$ *for functionals* $f : 1$

We are now in a position to clarify the problem left open in Howard's example 1.3.3 concerning the continuity of primitive recursive functionals.

1.5.27 Definition of continuity in NF

A functional Y of type $(o \to \tau) \to o$ is **continuous** (relative to the model NF), if for any argument $f : o \to \tau$, Yf depends only on finitely many values fi of f, formally:

$$NF \models \forall f^{o \to \tau} \exists x^o \forall g^{o \to \tau} \; f|x = g|x \to Yf = Yg$$

where $|$ is the restriction functional of 1.4.1.

1.5.28 Theorem: Continuity of primitive recursive functionals

For all types τ, the primitive recursive functionals $Y : (o \to \tau) \to o$ are continuous.

Proof. Let Y and f be normal primitive recursive functionals of types $(o \to \tau) \to o$ and $o \to \tau$, respectively. Without loss of generality, f is not the constant *suc*. By strong normalization 1.5.18, every reduction chain of Yf reduces Yf to the same numeral n. For the present proof, it suffices that Yf strictly reduces to n. We look at such a strict reduction chain

$$c_0 \equiv Yf, \; c_i \to_1 c_{i+1} \text{ for } i < k, \; c_k \equiv (Yf)^N \equiv n$$

and ask: what information about the primitive recursive functional f is needed for the correctness of the equations given by these k strict contractions which have the general form

$c[f, a[f]] \to_1 c[f, b[f]]$ where $a[f]$ *conv* $b[f]$ with $a[f]$ a minimal redex

There are two ways in which f can occur in the minimal redex $a[f]$:
(i) f is not at the head of $a[f]$. Then $a[f]$ starts with one of the constants of $L(T)$. In all these cases, it is easily seen that, for any $g : o \to \tau$, also $a[g]$ *conv* $b[g]$ and hence $c[g, a[g]] = c[g, b[g]]$.
(ii) $a[f]$ starts with f. Since f is normal and of type $o \to \tau$, $a[f]$ starts with ft for some functional $t : o$. But because $a[f]$ is a minimal redex, t is normal and hence a numeral. So, whenever $ft = gt$ holds for this particular numeral t, we have $a[g] = b[g]$ and hence $c[g, a[g]] = c[g, b[g]]$.
We now collect the finitely many numerals t which allow us to infer $c_i[g] = c_{i+1}[g]$ from $c_i[f] = c_{i+1}[f]$ for all $i < k$. Let m be a proper upper bound of these t. We then have:
If for all numerals $t < m$ $ft = gt$, then $c_i[f] = c_{i+1}[f]$ implies $c_i[g] = c_{i+1}[g]$ for all $i < k$ and therefore also $Yf = n$ implies $Yg = n$, i.e. $Yf = Yg$: Y is continuous.
For this proof, g does not even have to be a primitive recursive functional.

1.5.29   *Corollary: Non-primitive recursive functionals*

(1) *Equality functionals* $\mathsf{E}_{o\to\tau}$ *are not primitive recursive.*

(2) *There are no primitive recursive solutions* $Y^{(o\to\tau)\to o}$ *of*

$$(*) \qquad Yg^{o\to\tau} = 0 \leftrightarrow g = 0^{o\to\tau}$$

Proof of (2). Assume that there is a primitive recursive functional Y satisfying $(*)$. Then $Y0^{o\to\tau} = 0$; therefore, by the continuity theorem, in NF, $\exists x \forall g^{o\to\tau}\, 0^{o\to\tau}|x = g|x$ implies $Yg = 0$ which by $(*)$ in turn implies $g = 0^{o\to\tau}$ which is obviously not the case.

(2) implies (1), because if $\mathsf{E}_{o\to\tau}$ were primitive recursive, $\mathsf{E}_{o\to\tau}0^{o\to\tau}$ would be a primitive recursive solution Y of $(*)$, contradicting (2).

1.5.30   *Extending reductions to equality functionals*

If one wants to extend the reduction process to equality functionals, one has to define (E)-conversions for these non-primitive recursive functionals. Axioms (E) in 1.3.5, however, are not just equations like the defining axioms for the other functionals. Since conversions should be totally defined and cover the corresponding axioms at least for closed arguments, we define (E)-*conversions* as follows (cf. Troelstra [46]):

($E0$) $\mathsf{E}_\tau aa$ *conv*   0, if the term $a : \tau$ is in normal form.

($E1$) $\mathsf{E}_\tau ab$ *conv*   1, if a, b are distinct functionals of type τ in normal form.

These conversions extend the conversions formulated in 1.5.1 and lead to extended concepts of reduction, normalization etc. Certainly, the concepts of conversion and normality are now interdependent, but that is unproblematic, because the arguments a, b of an (E)-redex $\mathsf{E}ab$ contain less E than the redex itself. (E)-redexes, as defined here, are minimal. $\mathsf{E}ab$ is normal, if a, b are normal and not both closed, and not identical.

(E)-redexes extend parallel contraction only trivially, because they are minimal. With the exception of lemma 1.5.4 (6), the above arguments are independent of the theories T and $T_\wedge$ and extend without change to E-functionals down to the computability and the normalization theorems 1.5.17 and 18. These theorems clearly also hold for equality functionals, because for computable a, b, $\mathsf{E}ab$ strictly normalizes to 0, 1, or to a term $\mathsf{E}a^N b^N$ which is not closed. Therefore, proposition 1.5.20 holds also for I-T. (As bounded quantifiers are definable in I-T, we do not distinguish between I-T and I-$T_\wedge$ (cf. 1.3.6 and 1.3.7).)

We may then extend the type structure NF in 1.5.22 to a type structure $I\text{-}NF$ by adding all normal functionals containing an equality functional E_τ of higher type τ. Clearly, by our choice of (E)-conversions,

$$I\text{-}NF \models (Eab = 0 \leftrightarrow a = b) \wedge (Eab = 1 \leftrightarrow a \neq b)$$

so that by the argument for the soundness theorem 1.5.24, we have

1.5.31 *Soundness theorem for I-NF*

I-NF is a recursive model of I-T.

As above, this gives constructive meaning to the consistency statement for $I\text{-}T$ and its conservative extensions $I\text{-}HA^\omega$ and $I\text{-}HA^{\omega+}$. It is, however, not the case that $I\text{-}T$ proves all closed equations true in $I\text{-}NF$: The completeness statement for closed equations implicit in 1.5.4 (6) does not hold for $I\text{-}T$.

1.5.32 *Corollary: Incompleteness and undecidability of I-T*

There are functionals $t : o$ of $L(I\text{-}T)$ for which

$$I\text{-}T \not\vdash t = 0 \quad \text{as well as} \quad I\text{-}T \not\vdash t \neq 0$$

For functionals $t : o$ of $L(I\text{-}T)$, it is undecidable whether $I\text{-}T \vdash t = 1$.

As an example for incompleteness, take the extensionally equal functions $+0$ and I and form $t := E_1(+0)(I)$. By $(E1)$, t conv 1 and therefore, by soundness 1.5.31, $I\text{-}T \not\vdash t = 0$. But the full type structure $\mathbb{N}^\omega$, expanded by equality functionals, is also an extensional model of $I\text{-}T$ (with a non-recursive interpretation of E_1), and $\mathbb{N}^\omega \models t = 0$. So, $I\text{-}T \not\vdash t \neq 0$. The undecidability statement follows from (8) in 1.5.26, because

$$I\text{-}T \vdash E_1 f 0^1 = 1 \leftrightarrow f \neq 0^1$$

The incompleteness of $I\text{-}T$ even for closed equations of type o may be avoided by giving axiomatic status to the conversion rule $(E1)$, i.e. by extending $I\text{-}T$ by the axiom schema

$(E1)$ $E_\tau ab = 1$, if a, b are distinct functionals of type τ in normal form.

We doubt whether this axiom schema fits naturally into the axiomatic framework of $I\text{-}T$. It restores, however, the feature of T to $I\text{-}T + (E1)$ that conversions are (special cases of) axioms which makes 1.5.4 (6) valid for $I\text{-}T + (E1)$:

1.5.33   Corollary

(1) $I\text{-}NF$ is a recursive model of $I\text{-}T + (E1)$.

(2) For functionals a, b of $I\text{-}T$ of type τ,

$$I\text{-}T + (E1) \vdash a = b \Leftrightarrow a^N = b^N \Leftrightarrow I\text{-}NF \models a = b$$

(3) For functionals a, b of type τ, it is decidable whether $I\text{-}T + (E1) \vdash a = b$.

1.6   Modified realization and its hybrids

The Dialectica interpretation may be seen as the most reductive functional interpretation, since matrices A_D of Dialectica translated formulae belong to $L(T)$ and are as such strictly quantifier free - even free of bounded quantifiers. We now turn to the partly reductive - but still regular and idempotent - case, to modified realization mr, usually termed *modified realizability* in the literature. It puts universal quantifiers unchanged into the matrix of the translation A^{mr} of A such that the universal part of the prefix of A^{mr} remains empty. For that reason, many authors do not consider modified realization translations proper functional translations. However, formally they fit into the general concept of functional translation as defined in 0.1.15. Their associated hybrid translations are non-reductive, as the matrices of hybrid translated formulae contain arbitrary formulae of $L(HA^\omega)$.

Modified realization was introduced by Kreisel [29]. It parallels Kleene's [28] recursive realizability, using functionals of finite type instead of partial recursive functions. In this way, it combines Gödel's 1958 concept of functional interpretation with Kleene's 1945 concept of recursive realizability. Troelstra [45] and [46] extends this concept to modified q-realization mq which is related to Kleene's q-realizability in the same way in which modified realization is related to recursive realizability. q-realizability may in turn be seen as a hybrid of recursive realizability and a formalized version of Kleene's slash which itself is a metamathematical operation. Later, Troelstra [48] studies modified realization with truth mrt which again is related to Aczel's slash operation as mq is related to Kleene's slash.

Convention. If $A^{mr} \equiv \exists v\ A_{mr}[v]$, we follow the standard and write $v\ mr\ A$ - read: v modified realizes A - for $A_{mr}[v]$, and analogously for mq, mrt. If the tuple v is empty, it is often convenient to write $\emptyset\ mr\ A$ instead of $mr\ A$. In the context of modified realization, interpreting terms are called (modified) realizing terms, interpretable formulae and schemata are called (modified) realizable. - In this section, the bounded universal quantifier

$\forall x < t$ is - as in section 1.3 - considered defined by $\forall x (x < t \rightarrow \ldots)$ in HA^ω.

We start by defining **modified realization** mr, **modified q-realization** mq and **modified realization with truth** mrt.

1.6.1 *Recursive definition of modified realization translations mr, mq, and mrt*

To any formula A of $L(HA^\omega)$, mr assigns a formula $A^{mr} \quad \equiv \quad \exists v \; v \; mr \; A$ of $L(HA^\omega)$ as follows:

$$L(T)^{mr} \quad A^{mr} \equiv A \quad \text{for } A \in L(T)$$

Let A^{mr} be as above and $B^{mr} \equiv \exists y \; y \; mr \; B$, then

$$
\begin{aligned}
(\wedge)^{mr} \quad & (A \wedge B)^{mr} && \equiv \exists vy(v \; mr \; A \wedge y \; mr \; B) \\
(\rightarrow)^{mr} \quad & (A \rightarrow B)^{mr} && \equiv \exists Y \forall v(v \; mr \; A \rightarrow Yv \; mr \; B) \\
(\forall)^{mr} \quad & (\forall u A[u])^{mr} && \equiv \exists V \forall u \; Vu \; mr \; A[u] \\
(\exists)^{mr} \quad & (\exists u A[u])^{mr} && \equiv \exists uv \; v \; mr \; A[u]
\end{aligned}
$$

The recursive definition of mq runs as above, with mr replaced by mq and the following changes in the clauses $(\rightarrow)^{mr}$ and $(\exists)^{mr}$:

$$
\begin{aligned}
(\rightarrow)^{mq} \quad & (A \rightarrow B)^{mq} && \equiv \exists Y \forall v(A \wedge v \; mq \; A \rightarrow Yv \; mq \; B) \\
(\exists)^{mq} \quad & (\exists u A[u])^{mq} && \equiv \exists uv(A[u] \wedge v \; mq \; A[u])
\end{aligned}
$$

mrt is again defined recursively like mr, replacing mr by mrt and changing the clause $(\rightarrow)^{mr}$ into:

$$(\rightarrow)^{mrt} \quad (A \rightarrow B)^{mrt} \quad \equiv \exists Y((A \rightarrow B) \wedge \forall v(v \; mrt \; A \rightarrow Yv \; mrt \; B))$$

In the clauses for conjunction and implication, it may be assumed that not both components A, B are in $L(T)$.

Modified realization mr is obviously a regular functional translation, whereas mq and mrt are hybrid translations.

Translations of iterated existential quantification. For non-empty tuples u of variables, $(\exists)^{mr}$ and $(\exists)^{mrt}$ stay literally correct, whereas, for non-empty u, $(\exists)^{mq}$ is weakened to

$$HA^\omega \vdash (\exists u A[u])^{mq} \leftrightarrow \exists uv(A[u] \wedge v \; mq \; A[u])$$

This is seen by induction on the length of u.

The two hybrid translations mq and mrt are closely related:

1.6.2   *Proposition*

$$HA^\omega \vdash v \ mrt \ A \leftrightarrow (A \wedge v \ mq \ A)$$

Hence, $A \to A^{mq}$, $A \to A^{mrt}$, and $A \leftrightarrow A^{mrt}$ are equivalent in HA^ω.

Proof by subformula induction.

For $A \in L(T)$, $v \ mrt \ A \equiv A \equiv v \ mq \ A$.

($\wedge$) $v, y \ mrt \ (A \wedge B) \equiv v \ mrt \ A \wedge y \ mrt \ B$ is by I.H. equivalent to

$$(A \wedge v \ mq \ A) \wedge (B \wedge y \ mq \ B)$$

which is equivalent to $(A \wedge B) \wedge v, y \ mq \ (A \wedge B)$.

($\forall$) $V \ mrt \ \forall u A[u] \equiv \forall u \ Vu \ mrt \ A[u]$ is by I.H. equivalent to

$$\forall u(A[u] \wedge Vu \ mq \ A[u])$$

which is equivalent to $\forall u A[u] \wedge V \ mq \ \forall u A[u]$.

($\to$) $Y \ mrt \ (A \to B) \equiv (A \to B) \wedge \forall v(v \ mrt \ A \to Yv \ mrt \ B))$
is by I.H. equivalent to

$$(A \to B) \wedge \forall v((A \wedge v \ mq \ A) \to (B \wedge Yv \ mq \ B))$$

Due to the first conjunct $(A \to B)$, this is equivalent to

$$(A \to B) \wedge \forall v((A \wedge v \ mq \ A) \to Yv \ mq \ B) \equiv (A \to B) \wedge Y \ mq \ (A \to B)$$

($\exists$) $u, v \ mrt \ (\exists u A[u]) \equiv v \ mrt \ A[u])$ is by I.H. equivalent to
$A[u] \wedge v \ mq \ A[u]$ which is equivalent to

$$\exists u A[u] \wedge (A[u] \wedge v \ mq \ A[u]) \equiv \exists u A[u] \wedge u, v \ mq \ \exists u A[u]$$

There does not seem to exist an equally simple relation between mq and mr in HA^ω.

It is controversial whether this proposition gives the leading rôle to mq or to mrt. The proposition gives an outright definition of mrt in terms of mq which makes mq more fundamental. On the other hand, a tuple b of terms that mrt-realizes A, automatically also mq-realizes A, whereas, as we shall see, there are formulae A which are mq-realizable without being mrt-realizable in HA^ω. Technically, mrt is slightly easier to handle than mq, as its definition is closer to the definition of mr. We therefore introduce **modified G-realization** mrG for an arbitrary sentence G of $L(HA^\omega)$ in order to unify mr and mrt and the modified relization proofs for both of these translations. mrG-realization is essentially a special case of mr_P-realizability in Troelstra [46], Ch. III, section 4, which, however, is designed to unify mr and mq.

1.6.3 *Recursive definition of functional translations mrG*

Let G be a sentence of $L(HA^\omega)$. To any formula A of $L(HA^\omega)$, mrG assigns a formula $A^{mrG} \equiv \exists v \; v \; mrG \; A$ of $L(HA^\omega)$ as follows:

$$L(T)^{mrG} \quad A^{mrG} \equiv A \quad \text{for } A \in L(T)$$

Let A^{mrG} be as above and $B^{mrG} \equiv \exists y \; y \; mrG \; B$; then

$$(\wedge)^{mrG} \quad (A \wedge B)^{mrG} \equiv \exists vy(v \; mrG \; A \wedge y \; mrG \; B)$$
$$(\rightarrow)^{mrG} \quad (A \rightarrow B)^{mrG} \equiv \exists Y((G \rightarrow A \rightarrow B) \wedge \forall v(v \; mrG \; A \rightarrow Yv \; mrG \; B))$$
$$(\forall)^{mrG} \quad (\forall u A[u])^{mrG} \equiv \exists V \forall u(Vu \; mrG \; A[u])$$
$$(\exists)^{mrG} \quad (\exists u A[u])^{mrG} \equiv \exists uv \; v \; mrG \; A[u]$$

The only difference to the definition of mrt is the occurrence of G in $(\rightarrow)^{mrG}$, replacing the conjunct $(A \rightarrow B)$ by $G \rightarrow (A \rightarrow B)$. Two special cases are of interest for us:

(i) $G \equiv \bot$. In this case, the conjunct $\bot \rightarrow (A \rightarrow B)$ is trivially true and may be cancelled altogether, in particular as it does not contribute to the prefix. By this cancellation, $mr\bot$ becomes mr.

(ii) $G \equiv \top$. In this case, the antecedent $\top$ in $\top \rightarrow (A \rightarrow B)$ may be cancelled, and $mr\top$ becomes mrt.

So, whatever holds for mrG in general, holds equally for mr and for mrt. We do not know of any applications of mrG for other sentences, e.g. Gödel sentences G.

These translations, including mq, are designed to extract positive existential information contained in A and leave universal data unchanged. The translations therefore have almost no effect on $\exists$free formulae:

1.6.4 *Lemma*

If I is one of the translations mr, mq, mrt, mrG and A is $\exists$free, then

$$A^I \equiv \emptyset \; I \; A \text{ and } HA^\omega \vdash A^I \leftrightarrow A$$

For mr, even $A^{mr} \equiv A$ for $\exists$free A. Moreover, $v \; mr \; A$ is $\exists$free for any formula A; hence mr is idempotent.

Proof by subformula induction. The statements on mr are seen by running through the definition of mr. For mrG – which includes the statement for mrt – and for mq, we show only the step for implication. For $\exists$free A, B, by $(\rightarrow)^{mrG}$ and I.H.

$$(A \rightarrow B)^{mrG} \equiv (G \rightarrow A \rightarrow B) \wedge (\emptyset \; mrG \; A \rightarrow \emptyset \; mrG \; B)$$

which is by I.H. equivalent to $(G \to A \to B) \wedge (A \to B)$, i.e. to $A \to B$. Similarly, by $(\to)^{mq}$ and I.H.

$$(A \to B)^{mq} \equiv (A \wedge \emptyset \, mq \, A \to \emptyset \, mq \, B)$$

which is by I.H. equivalent to $(A \wedge A \to B)$, i.e. to $(A \to B)$.

Corollary. For translations I as above, tuples u of variables, and ∃free $A[u]$,

$$HA^\omega \vdash (\exists u A[u])^I \leftrightarrow \exists u A[u]$$

Proof. For mr, we again have identity; for mrG, the corollary is immediate; for mq, by the remark on iterated existential quantification in 1.6.1 above,

$$(\exists u A[u])^{mq} \equiv \exists u (A[u] \wedge \emptyset \, mq \, A[u]) \leftrightarrow \exists u (A[u] \wedge A[u]) \leftrightarrow \exists u A[u]$$

As mentioned above, under hypothesis G, the definition of mrG reduces to the definition of mrt:

1.6.5 *Lemma*

$$HA^\omega \vdash G \to (v \, mrG \, A \leftrightarrow v \, mrtA)$$

The proof by subformula induction is routine.

Corollary. $HA^\omega \vdash v \, mrG \, A \to (G \to A)$

This follows from the lemma and proposition 1.6.2. As part of the definition of mrG, we have:

1.6.6 *Lemma*

If in an extension H of HA^ω,

$$H \vdash (G \to C \to D) \wedge v \, mrG \, C \to v \, mrG \, D,$$

then the tuple $\lambda v.v \;\; mrG$-realizes $C \to D$ in H.

1.6.7 *Modified realization theorem for mrG*

$$HA^\omega \overset{mrG}{\hookrightarrow} HA^\omega$$

More precisely: If $HA^\omega \vdash A$, then there is a tuple b of terms with $FV(b) \subset FV(A)$ such that for all sentences G:

$$HA^\omega \vdash b \, mrG \, A$$

Proof by induction on deductions. Many axiom schemata and inference rules (A1, A3 to A⊥, Q1 to Q4) consist exclusively of implications $A \to B$. In these cases, their mrG-translation falls into two parts: The one - which we may call the G-part for the moment - consists entirely of implications $G \to (A \to B)$, and the G-part mimicks the given axiom or inference, maintaining validity, essentially due to Frege's rule. Therefore, this G-part does not need detailed discussion. The other part - let us call it the mr-part - consists of formulae $\forall v(v \; mrG \; A \to Y v \; mrG \; B)$, and this part needs our attention.

A1 and similarly A4, apply lemma 1.6.6: $\lambda v.v$ and $\lambda vy.v$ respectively are the mrG-realizing term tuples.

A2. By I.H., there are $v_0, Y \; mrG$-realizing A and $A \to B$ which implies $v_0 \; mrG \; A$ and $\forall v(v \; mrG \; A \to Y v \; mrG \; B)$. So, $Y v_0 \; mrG$-realizes B.

A3. By I.H., there are Y, Z which mrG-realize $A \to B$, $B \to C$ respectively, hence

$$\forall v(v \; mrG \; A \to Y v \; mrG \; B) \text{ and } \forall y(y \; mrG \; B \to Z y \; mrG \; C)$$

From this we get

$$\forall v(v \; mrG \; A \to Z(Y v) \; mrG \; C)$$

Since the G-part yields $G \to A \to C$, composition $Z \circ Y := \lambda v.Z(Y v)$ mrG-realizes $A \to C$.

A5. By I.H., there are terms $Y \; mrG$-realizing $A \wedge B \to C$, hence

$$\forall vw(v \; mrG \; A \wedge w \; mrG \; B \to Y vw \; mrG \; C)$$

This implies (by A5, among other laws)

$$\forall v(v \; mrG \; A \to \forall w(w \; mrG \; B \to Y vw \; mrG \; C))$$

By the corollary to 1.6.5, $A \to B \to C$ implies $v \; mrG \; A \to (G \to B \to C)$. Thus we obtain

$$\forall v(v \; mrG \; A \to ((G \to B \to C) \wedge \forall w(w \; mrG \; B \to Y vw \; mrG \; C)))$$

which together with $G \to A \to B \to C$ is $Y \; mrG \; (A \to B \to C)$.

A6 is shown by reading the proof for A5 backwards, but without recourse to 1.6.5, simply cancelling the conjunct $G \to B \to C$.

A7. By I.H., we have Y_0, Y_1 which mrG-realize $A \to B$ and $A \to C$ respectively, so

$$\forall v(v \; mrG \; A \to (Y_0 v \; mrG \; B \wedge Y_1 v \; mrG \; C)$$

Since we have $G \to A \to (B \wedge C)$, the combined tuple Y_0, Y_1 also mrG–
realizes $A \to (B \wedge C)$.

$A\perp$ is realized, e.g., by 0-functionals.

Axioms (K), (S), (suc), (R0), (Rsuc) and, in the presence of product types,
(P) are mrG-realized trivially.

(Eq) By I.H., there are terms Y such that

$$\forall v(v \, mrG \, A \to a = b) \text{ and } Y \, mrG \, F[a]$$

By (Eq), this implies $\forall v(v \, mrG \, A \to Y \, mrG \, F[b])$. Since $G \to A \to F[b]$,
Y also mrG-realizes $A \to F[b]$.

We come to the induction rule (IND). By I.H., there are terms v_0 and $Y[x]$
such that

$$v_0 \, mrG \, A[0] \text{ and } \forall v(v \, mrG \, A[x] \to Y[x]v \, mrG \, A[suc\,x])$$

By simultaneous recursion, we define a term tuple f by

$$f0 = v_0 \text{ and } f(suc\,x) = Y[x](fx)$$

By (IND), we obtain $ft \, mrG \, A[t]$.

Q1. Since $v \, mrG \, \forall u A[u] \to (vb) \, mrG \, A[b]$, the term tuple $\lambda v.vb \, mrG$-
realizes Q1.

Q2. Since $v \, mrG \, A[b] \to b, v \, mrG \, \exists u A[u]$, the tuple $\lambda v.b, \lambda v.v \, mrG$-realizes
Q2.

Q3. By I.H., there are terms $Y \equiv Y[u]$ realizing $A \to B[u]$. So, by Q3,

$$\forall v(v \, mrG \, A \to \forall u Y v \, mrG \, B[u]) \equiv \forall v(v \, mrG \, A \to \lambda u.Y v \, mrG \, \forall u B[u])$$

That is, the tuple $\lambda vu.(Y[u]v) \, mrG$-realizes $A \to \forall u B[u]$.

Q4. By I.H., there are terms $Y \equiv Y[u] \, mrG$-realizing $A[u] \to B$. So, by
Q4,

$$\forall v(v \, mrG \, A[u] \to Y v \, mrG \, B) \equiv \forall uv(u, v \, mrG \, \exists u A[u] \to Y v \, mrG \, B)$$

Hence, here the tuple $\lambda uv.(Y[u]v) \equiv \lambda u.Y[u] \, mrG$-realizes $\exists u A[u] \to B$.

Comment. Given a derivation in HA^ω, our proof on the one hand re-
produces that derivation under the additional antecedent G; on the other,
it produces a derivation along the corresponding mrG-translations which
allows a computation of mrG-realizing terms. Both sides do not inter-
fere with another, with one exception, viz. inferences A5 which gener-
ate an iterated implication $A \to B \to C$ out of a simple implication
$A \wedge B \to C$. In this situation, we have to derive an additional mixed
implication $v \, mrG \, A \to (G \to B \to C)$ using corollary 1.6.5.

As a corollary, we obtain realization theorems for Kreisel's and Troelstra's
concepts of modified realization introduced above.

1.6.8 Uniform modified realization theorem for mr, mrt, and mq

If $HA^\omega \vdash A$, then there is a tuple b of terms with $FV(b) \subset FV(A)$ such that:
$$HA^\omega \vdash b \ mr \ A \wedge b \ mrt \ A \wedge b \ mq \ A \quad \text{hence:}$$
$$HA^\omega \overset{mr}{\hookrightarrow} HA^\omega \qquad HA^\omega \overset{mrt}{\hookrightarrow} HA^\omega \qquad HA^\omega \overset{mq}{\hookrightarrow} HA^\omega$$

Proof. For mr, set $G \equiv \bot$ in 1.6.7. By (i) in 1.6.3, $mr\bot$ reduces to mr. For mrt, set $G \equiv \top$. By remark (ii) in 1.6.3, $mr\top$ reduces to mrt. For mq, the theorem follows from the theorem for mrt by proposition 1.6.2.

Since the translation mr is regular and idempotent, applying lemma 0.1.19, we immediately obtain a stronger result for mr:

1.6.9 Corollary

$$HA^\omega + \{A^{mr} \leftrightarrow A\} \overset{mr}{\hookrightarrow} HA^\omega,$$
and $HA^\omega + \{A^{mr} \leftrightarrow A\}$ is mr-maximal relative to HA^ω:
Any formula modified realizable in HA^ω is derivable in $HA^\omega + \{A^{mr} \leftrightarrow A\}$.

This argument does not transfer to the hybrid translations mrt or mq. It is yet our aim to characterize all translations mr, mrt, mq, and mrG. We first collect data concerning the general translation mrG, as far as that is easily possible, and add data completing the characterization of mr. The following axioms of choice and independence of premise are adequate and as weak as possible for this purpose:

1.6.10 ∃free choice and independence of premise, the theory $HA^{\omega\top}$

$AC_{\exists free} \qquad \forall x \exists y A[x, y] \to \exists Y \forall x A[x, Yx]$
with $A[x, y]$ an ∃free formula and x, y tuples of variables of arbitrary type.

$IP_{\exists free} \qquad (A \to \exists y B[y]) \to \exists y(A \to B[y])$
with $A, B[y]$ ∃free formulae and y a tuple of variables of arbitrary type.
$$HA^{\omega\top} :\equiv HA^\omega + AC_{\exists free} + IP_{\exists free}$$

$AC_{\exists free}$ is a subschema of AC, whereas $IP_{\exists free}$ is more general than $IP_\wedge$ concerning the antecedent, and less general concerning the consequent. In the presence of product types, our formulations of $AC_{\exists free}$ and $IP_{\exists free}$ are obviously equivalent to their restrictions to a single variable y. We do not know whether this is true in the absence of product types.

1.6.11   Lemma

1. HA^ω $\vdash A^{mrG}$ $\leftrightarrow A$ *for* $A \in L(T)$
2. HA^ω $\vdash (A \wedge B)^{mrG}$ $\leftrightarrow (A^{mrG} \wedge B^{mrG})$
3. HA^ω $\vdash (A \to B)^{mrG}$ $\to (G \to A \to B) \wedge (A^{mrG} \to B^{mrG})$
4. HA^ω $\vdash (\forall u A[u])^{mrG}$ $\to \forall u (A[u]^{mrG})$
5. $HA^\omega + AC$ $\vdash \forall u (A[u]^{mrG})$ $\to (\forall u A[u])^{mrG}$
6. HA^ω $\vdash (\exists u A[u])^{mrG}$ $\leftrightarrow \exists u (A[u]^{mrG})$
7. $HA^{\omega\top}$ $\vdash (A^{mr} \to B^{mr}) \to (A \to B)^{mr}$
8. $HA^\omega + AC_{\exists free} \vdash$ $\forall u (A[u]^{mr}) \to (\forall u A[u])^{mr}$

Proof. 1. and 6. are syntactic identities. 2. to 5. are a simple exercise in intuitionistic logic and are left to the reader.

7. Since $v \, mr \, A$ and $y \, mr \, B$ are $\exists$free,

$\vdash \exists v(v \, mr \, A) \to \exists y(y \, mr \, B) \leftrightarrow \forall v(v \, mr \, A \to \exists y(y \, mr \, B))$
           $\to \forall v \exists y(v \, mr \, A \to y \, mr \, B)$ by $IP_{\exists free}$
           $\to \exists Y \forall v(v \, mr \, A \to Yv \, mr \, B)$ by $AC_{\exists free}$

which is $\vdash (A^{mr} \to B^{mr}) \to (A \to B)^{mr}$.

8. Since $v \, mr \, A[u]$ is $\exists$free,

$\vdash \forall u \exists v(v \, mr \, A[u]) \to \exists V \forall u(Vu \, mr \, A[u])$ by $AC_{\exists free}$

which is $\forall u (A[u]^{mr}) \to (\forall u A[u])^{mr}$.

For $mr \equiv mr\bot$, i.e. for $G \equiv \bot$, this lemma contains the induction steps for showing by subformula induction:

1.6.12   Proposition

For all formulae C of $L(HA^\omega)$:

$$HA^\omega + AC_{\exists free} + IP_{\exists free} \vdash C^{mr} \leftrightarrow C$$

In the other direction, we can prove more, even in the general setting of mrG-realization. The prefix of a translated formula A^{mrG} collects information on the strictly positive occurrences of existential quantifiers in A. Formulae A without such strictly positive occurrences of $\exists$ therefore have a mrG-translation $\emptyset \, mrG \, A$. This class of formulae has first been studied by Harrop.

1.6.13   Recursive definition of Harrop formulae

1. Any formula of $L(T)$ is Harrop.
2. If A and B are Harrop, then so are $A \wedge B$ and $\forall u A$.
3. If B is Harrop, then, for any formula A, $A \to B$ is Harrop.

Though the class of Harrop formulae is larger than the class of ∃free formulae, there is still a simple relation between Harrop formulae and their *mrG*-translation, though weaker than the information on ∃free formulae in 1.6.4:

1.6.14   Lemma

For Harrop formulae A, $A^{mrG} \equiv \emptyset\ mrG\ A$ *and*

$$HA^\omega \vdash G \to (A \to \emptyset\ mrG\ A)$$

Proof by subformula induction. We show the induction step for implications $A \to B$, the only nontrivial case. We have $G \to (v\ mrG\ A \to A)$ by corollary 1.6.5 and $G \to (B \to \emptyset\ mrG\ B)$ by I.H.. Hence $G \wedge (A \to B)$ implies

$$(G \to A \to B) \wedge \forall v(v\ mrG\ A \to \emptyset\ mrG\ B) \equiv \emptyset\ mrG\ (A \to B)$$

Corollary. *For A Harrop,*

$$HA^\omega \vdash A \to \emptyset\ mq\ A$$

This follows from the lemma for $G \equiv \top$ by proposition 1.6.2.

1.6.15   *Schema of independence of Harrop premise*

$$IP_H \quad (A \to \exists y B[y]) \to \exists y (A \to B[y])$$

where A is a Harrop formula and $B[y]$ is an arbitrary formula.

We now obtain a converse of proposition 1.6.12 in stronger form:

1.6.16   *Proposition*

In $HA^\omega + \{C^{mrG} \leftrightarrow C\}$, *the schemata AC and* IP_H *are derivable.*

Proof.

$$(\forall x \exists y A[x,y])^{mrG} \equiv \exists Y V \forall x(V x\ mrG\ A[x, Yx]) \equiv (\exists Y \forall x A[x, Yx])^{mrG} :$$

antecedent and consequent of axioms AC have identical mrG-translations. Therefore, AC follows from two axioms $C^{mrG} \leftrightarrow C$.
For A Harrop, by lemma 1.6.14, $(A \to \exists y B[y])^{mrG}$ implies $\exists y, y'(\emptyset\ mrG\ A \to y'\ mrG\ B[y])$ and in turn $\exists y(A^{mrG} \to B[y]^{mrG})$. Thus, three axioms $C^{mrG} \leftrightarrow C$ yield IP_H.

Since AC contains $AC_{\exists free}$, and IP_H contains $IP_{\exists free}$, we obtain as a corollary to propositions 1.6.12 and 1.6.16 for $mr = mr\bot$:

1.6.17 Characterization and extended realization theorem for mr

The theories

$$HA^\omega + \{A^{mr} \leftrightarrow A\}$$
$$HA^\omega + AC_{\exists free} + IP_{\exists free} \equiv HA^{\omega\top}$$
$$HA^\omega + AC + IP_H$$

are equivalent. They are mr-realizable in HA^ω and mr-maximal relative to HA^ω.

The last line follows by corollary 1.6.9. These theories are in fact *mr*-realizable in the ∃free fragment of HA^ω and therefore also conservative over this fragment, as may be seen from the proof of theorem 1.6.7.

In contrast to regular functional interpretations, hybrid interpretations like *mrt* and *mq* serve to show closure of theories like HA^ω under metamathematically relevant rules. We apply *mrt*-realizability (*mq*-realizability would serve the same purpose) to show explicit definability in HA^ω and closure of HA^ω under rules corresponding to the axiom schemata AC and IP_H.

1.6.18 Theorem: Admissible rules

HA^ω is closed under explicit definability and under explicit forms of the rule of choice and the rule of independence of Harrop premise:

(ED)   *If $\vdash \exists y B[y]$, then there are terms c such that $\vdash B[c]$*
(Rule-AC)   *If $\vdash \forall x^\sigma \exists y^\tau A[x,y]$ then $\vdash A[x, fx]$ for a term $f : \sigma \to \tau$.*
(Rule-IP_H) *If $\vdash A \to \exists y B[y]$ and A is Harrop, then there are terms c such that $\vdash A \to B[c]$*

Proof. (ED) is the special case $A \equiv \top$ of *(Rule-IP_H)*. *(Rule-AC)* follows from (ED) by defining $f := \lambda x.c[x]$. It remains to show *(Rule-IP_H)*. Let $HA^\omega \vdash A \to \exists y B[y]$ for a Harrop formula A. By *mrt*-realization 1.6.8, there are terms c, Y such that $HA^\omega \vdash c, Y \; mrt \; (A \to \exists y B[y])$, i.e.

$$HA^\omega \vdash (A \to \exists y B[y]) \wedge (\emptyset \; mrt \; A \to Y \; mrt \; B[c])$$

The first conjunct is redundant, and since A is Harrop, $A \to \emptyset \; mrt \; A$ by lemma 1.6.14 and $Y \; mrt \; B[c] \to B[c]$ by proposition 1.6.2. Therefore, as claimed,

$$HA^\omega \vdash A \to B[c]$$

It should be noticed that these admissible rules are rules *in parameters*, in contrast to the corresponding *closed* rules which are only special cases of the rules presented here. In *(ED)*, for example, the formula $\exists y B[y]$ may contain free variables which then may also occur in the exemplifying terms c. The closed version (ED^c) applies only to sentences $\exists y B[y]$.

It remains to extend the results obtained for mr in theorem 1.6.17 to the hybrid translations mrt, mq, and mrG. We keep to our unifying approach and look at mrG again.

1.6.19 Definition of the theory $HA^{\omega G}$

For a class Δ of formulae, put $(G \to \Delta) := \{G \to D \mid D \in \Delta\}$ and
$$HA^{\omega G} :\equiv HA^{\omega} + (G \to AC_{\exists free}) + (G \to IP_{\exists free})$$
Thus, $HA^{\omega \perp} \equiv HA^{\omega}$ and $HA^{\omega \top} \equiv HA^{\omega} + AC_{\exists free} + IP_{\exists free}$. This last abbreviation has already been introduced in definition 1.6.10. With this notation, mrG-realization 1.6.7 extends to:

1.6.20 Extended mrG-realization theorem

$$HA^{\omega} + AC_{\exists free} + IP_{\exists free} \overset{mrG}{\hookrightarrow} HA^{\omega G}$$

Proof. Extending the proof of 1.6.8, we only have to mrG-realize $AC_{\exists free}$ and $IP_{\exists free}$ in $HA^{\omega G}$.

mrG-realization of $AC_{\exists free}$. For $\exists$free A, for $C :\equiv \forall x \exists y A[x,y]$ and $D :\equiv \exists Y \forall x\, A[x, Yx]$, we have
$$Y\ mrG\ C \equiv \forall x\ \emptyset\ mrG\ A[x, Yx] \equiv Y\ mrG\ D$$
Since $G \to C \to D$ is an axiom $(G \to AC_{\exists free})$ of $HA^{\omega G}$, $\lambda Y.Y\ mrG$-realizes $AC_{\exists free}$ in $HA^{\omega G}$ by lemma 1.6.6.

mrG-realization of $IP_{\exists free}$. For $\exists$free A, B, for $C \equiv A \to \exists y B[y]$ and $D \equiv \exists y(A \to B[y])$,
$$y\ mrG\ C \equiv (G \to C) \wedge (\emptyset\ mrG\ A \to \emptyset\ mrG\ B[y]) \quad \text{and}$$
$$y\ mrG\ D \equiv (G \to A \to B[y]) \wedge (\emptyset\ mrG\ A \to \emptyset\ mrG\ B[y])$$
The right conjuncts $F[y] := (\emptyset\ mrG\ A \to \emptyset\ mrG\ B[y])$ of $y\ mrG\ C$ and $y\ mrG\ D$ are identical, and $F[y]$ also implies the left conjunct $(G \to A \to B[y])$ of $y\ mrG\ D$, because
$\vdash G \to A \to \emptyset\ mrG\ A$ by lemma 1.6.14 and
$\vdash G \to \emptyset\ mrG\ B[y] \to B[y]$ by lemma 1.6.5. Therefore
$$\vdash y\ mrG\ C \to y\ mrG\ D$$
Since $G \to C \to D$ is an axiom $G \to IP_{\exists free}$ of $HA^{\omega G}$, $\lambda y.y\ mrG$-realizes $C \to D$ in $HA^{\omega G}$ by lemma 1.6.6.

As special cases, we obtain in the same way as corollary 1.6.9 is obtained from theorem 1.6.8:

1.6.21   *Extended uniform realization theorem*

$$HA^\omega + AC_{\exists free} + IP_{\exists free} \overset{mr}{\hookrightarrow} HA^\omega$$

$$HA^\omega + AC_{\exists free} + IP_{\exists free} \overset{mrt}{\hookrightarrow} HA^\omega + AC_{\exists free} + IP_{\exists free}$$

$$HA^\omega + AC_{\exists free} + IP_{\exists free} \overset{mq}{\hookrightarrow} HA^\omega + AC_{\exists free} + IP_{\exists free}$$

Again, the same terms realize the same theorems under all three translations.

1.6.22   *Remarks*

(i) Because mr is regular and idempotent, extended mr-realization is, by lemma 0.1.19, already a consequence of mr-realization 1.6.9 and mr-characterization 1.6.17 or proposition 1.6.16. Such an argument does not work for mrG in general nor for mq.

(ii) Since v mr A is $\exists$free, it is easily seen that $HA^{\omega\top}$ is mr-realized in the $\exists$free fragment of HA^ω. As a corollary to extended mr-realization, we may therefore note:
$HA^\omega + AC + IP_H$ *is a conservative extension of the $\exists$free fragment of HA^ω and hence also of $T_\wedge$.*
This result parallels the conservativity of $HA^{\omega+}$ over $T_\wedge$, but neither result is a corollary to the other.

(iii) We leave it to the reader to verify that axioms AC and IP_H are mq-realized already in HA^ω, in contrast to the situation for mrt, but in accordance with proposition 1.6.2. This does not mean that extended mq-realization can be strengthened beyond theorem 1.6.21. The point is that the class of formulae mq-realized in HA^ω is not deductively closed:
Let A be an instance of IP_H or AC underivable in HA^ω. As mentioned, A is mq-realized in HA^ω and hence $HA^\omega \vdash \exists v\, v$ mq A. For a variable u not in A, $(\exists u A)^{mq} \equiv \exists u v (A \wedge v$ mq $A)$. Since this implies A, $\exists u A$ is not mq-realized in HA^ω, but only in $HA^{\omega\top}$, in spite of $HA^\omega \vdash A \leftrightarrow \exists u A$.

We are now in a position to characterize the translation mrG for arbitrary G.

1.6.23   *Proposition*

$$HA^\omega + AC_{\exists free} + IP_{\exists free} \vdash A^{mrG} \leftrightarrow A$$

Proof. By the characterization theorem for mr 1.6.17,

(1) $HA^\omega + AC_{\exists free} + IP_{\exists free} \vdash A^{mr} \leftrightarrow A$

By extended mrG-realization 1.6.20, $A^{mr} \leftrightarrow A$ is therefore mrG-realized in $HA^{\omega G}$ and a fortiori in $HA^{\omega \top}$. This implies

(2) $HA^\omega + AC_{\exists free} + IP_{\exists free} \vdash (A^{mr} \leftrightarrow A)^{mrG}$

so that by lemma 1.6.11,3

(3) $HA^\omega + AC_{\exists free} + IP_{\exists free} \vdash (A^{mr})^{mrG} \leftrightarrow A^{mrG}$

But $A^{mr} \equiv \exists v A_{mr}$ with $\exists$free matrix A_{mr}. Therefore, by the corollary to 1.6.4,

(4) $HA^\omega \vdash (A^{mr})^{mrG} \leftrightarrow A^{mr}$

(1), (4), and (3) taken together yield the proposition.

Remark: Normal forms. In this proof, the formula A^{mr} plays a key rôle, though it does not occur in the proposition. The reason is the following:
(i) A^{mr} is $\exists$-*prenex*, i.e. it is of a form $\exists v B$ with $\exists$free B, and is therefore easily mrG-translated, as shown in (4) above.
(ii) Within the relevant theory $HA^{\omega \top}$, A^{mr} is, by (1) above, even an $\exists$-*prenex normal form* of A. Since this theory is mrG-realizable (cf. (2) and (3)), it suffices to work with these $\exists$-prenex normal forms instead of with arbitrary formulae.
It will be seen that the consideration of appropriate normal forms will be useful for the characterization of other hybrid or classical translations, too.

1.6.24   *Proposition*

$$HA^\omega + AC_{\exists free} + IP_{\exists free} \vdash A^{mq} \leftrightarrow A$$

Proof. By proposition 1.6.2, the implication $A \to A^{mq}$ is equivalent to $A^{mrt} \leftrightarrow A$, and its derivability is therefore equivalent to the case $G \equiv \top$ of the previous proposition. The implication $A^{mq} \to A$ follows by subformula induction. We show the cases for implication and existential quantification. $(A \to B)^{mq}$ implies $(A \wedge A^{mq}) \to B^{mq}$. But $A \to (A \wedge A^{mq})$ by the implication already shown, and $B^{mq} \to B$ by I.H. Hence, $(A \to B)^{mq} \to (A \to B)$.
$(\exists u A[u])^{mq}$ obviously implies $\exists u A[u]$.

Putting these results together, we obtain:

1.6.25   Characterization theorem for mrG and mq

For any sentence G, the theories

$$HA^\omega + \{A^{mr} \leftrightarrow A\}$$
$$HA^\omega + \{A^{mrt} \leftrightarrow A\}$$
$$HA^\omega + \{A^{mq} \leftrightarrow A\}$$
$$HA^\omega + \{A^{mrG} \leftrightarrow A\}$$
$$HA^\omega + AC_{\exists free} + IP_{\exists free} \equiv HA^{\omega\top}$$
$$HA^\omega + AC + IP_H$$

are equivalent.

Proof. By the characterization theorem 1.6.17 for mr, the theories in the first and in the last two lines are equivalent. Therefore and by propositions 1.6.23 and 1.6.24, $HA^{\omega\top}$ proves at least as much as any theory in the list. On the other hand, by proposition 1.6.16, for any sentence G, $HA^\omega + \{A^{mrG} \leftrightarrow A\}$ proves $AC + IP_H$. Finally, by proposition 1.6.2, already $HA^\omega + \{A \to A^{mq}\}$ proves $A^{mrt} \leftrightarrow A$ and therefore, by proposition 1.6.16 for $G \equiv \top$, again $AC + IP_H$.

1.7   Hybrids of the $\wedge$-interpretation

At this point, it is a natural question to ask whether reductive functional interpretations like the Dialectica and the $\wedge$-interpretation possess hybrids similar to the hybrids mrt and mq of modified realization mr. It is immediate that the Dialectica interpretation does not possess hybrids, because for the Dialectica interpretation of rule A7, the matrices of the translated formulae have to possess characteristic terms in T which is not the case for hybrid translations which carry along formulae of arbitrary quantifier complexity in their matrices.

The $\wedge$-interpretation, however, does possess hybrids. We look at three hybrid translations the first two of which, $\wedge q$ and $\wedge t'$, are defined in complete analogy to mq and mrt respectively. The first, $\wedge q$, was introduced and studied by M. Stein [39] and later rediscovered by Jørgensen [27]. The second, $\wedge t'$, does not yield a functional interpretation. Here, the analogy to mrt is misleading.

A more systematic path to construct a hybrid with truth for the $\wedge$-translation is followed by Gaspar and Oliva [20] by adjoining conjunctively the formula to be translated to the matrix of its translation, in each clause of its recursive definition in which it is not obviously superfluous. With their version of the $\wedge$-translation, this means adjoining A in case A is an implication or a universal formula. However, in the clause for implication, they use an inhabited bound equivalent to $\forall x \leq Xvz$ in contrast to the formulation $\forall x < Xvz$ of the $\wedge$-translation in definition 1.2.5. By an analysis where the inhabited bound is essential for their translation, a hybrid with truth $\wedge t$ of $\wedge$ itself is constructed as the third hybrid translation under consideration: In the $\wedge t$-translation of implication $A \to B$, an additional copy of A is needed as a conjunct to the antecedent $\forall x < Xvz\, A_{\wedge t}[v, Wxvz]$. We describe all three tranlations under recourse to definition 1.2.5 of $\wedge$.

1.7.1 Recursive definition of functional translations $\wedge q$, $\wedge t'$, and $\wedge t$ of HA^ω

(1) $\wedge q$ is defined like $\wedge$ in 1.2.5, with $\wedge$ replaced by $\wedge q$ and the following changes in the clauses $(\to)^\wedge$ and $(\exists)^\wedge$:

$(\to)^{\wedge q}$ $(A \to B)^{\wedge q} \equiv \exists XWY \forall vz (A \wedge \forall x < Xvz A_{\wedge q}[v, Wxvz] \to B_{\wedge q}[Yv, z])$
in case the tuple w is not empty

$(\to)_0^{\wedge q}$ $(A \to B)^{\wedge q} \equiv \exists Y \forall vz (A \wedge A_{\wedge q}[v] \to B_{\wedge q}[Yv, z])$ for empty w

$(\exists)^{\wedge q}$ $(\exists u A[u])^{\wedge q} \equiv \exists uv \forall w (A[u] \wedge A_{\wedge q}[u, v, w])$

(2) $\wedge t'$ is defined recursively like $\wedge$, replacing $\wedge$ by $\wedge t'$ and changing the clause $(\to)^\wedge$ into:

$(\to)^{\wedge t'}$ $(A \to B)^{\wedge t'} \equiv \exists XWY \forall vz ((A \to B) \wedge$
$(\forall x < Xvz\, A_{\wedge t'}[v, Wxvz] \to B_{\wedge t'}[Yv, z]))$
in case the tuple w is not empty

$(\to)_0^{\wedge t'}$ $(A \to B)^{\wedge t'} \equiv \exists Y \forall vz ((A \to B) \wedge (A_{\wedge t'}[v] \to B_{\wedge t'}[Yv, z]))$ for empty w

(3) $\wedge t$ is again defined recursively like $\wedge$, replacing $\wedge$ by $\wedge t$ and changing the clauses $(\to)^\wedge$ and $(\forall)^\wedge$ into:

$(\to)^{\wedge t}$ $(A \to B)^{\wedge t} \equiv \exists XWY \forall vz ((A \to B) \wedge$
$(A \wedge \forall x < Xvz\, A_{\wedge t}[v, Wxvz] \to B_{\wedge t}[Yv, z]))$
in case the tuple w is not empty

$(\to)_0^{\wedge t}$ $(A \to B)^{\wedge t} \equiv \exists Y \forall vz ((A \to B) \wedge (A_{\wedge t}[v] \to B_{\wedge t}[Yv, z]))$ for empty w

$(\forall)^{\wedge t}$ $(\forall u\, A[u])^{\wedge t} \equiv \exists V \forall uw (\forall u A[u] \wedge A_{\wedge t}[u, Vu, w])$

For disjunction as defined in 1.2.2, we get up to logical equivalence for $I = \wedge q, \wedge t', \wedge t$

$(\vee)^I$ $(A \vee B)^I \equiv \exists xvy \forall wz ((x = 0 \to A \wedge A_I) \wedge (x \neq 0 \to B \wedge B_I))$

As with modified realization and its hybrids, the translations $\wedge$, $\wedge q$, $\wedge t'$, and $\wedge t$ produce equal prefixes for a given formula. By an easy subformula induction, we also see in partial analogy to proposition 1.6.2:

$$HA^\omega \vdash \forall w A_{\wedge t'}[v, w] \to A$$

However, this does not suffice for a $\wedge t'$-interpretation of HA^ω. To the contrary:

1.7.2   *Proposition*

There is a theorem of HA^ω – even a tautology of HA – which is not $\wedge t'$-interpretable in HA^ω.

Proof. A hint to this defect is contained in the proof of mrt-realizability. At one single point, in the induction step for A5, an mrt-translated subformula $v\ mrt\ A$ enters into the untranslated part, because the conclusion of A5 is an iterated implication. Our counterexample will be of that form.

Let $con\ z^o$ express that z^o is not the code of a proof of $\bot$ in HA^ω. Then con is a primitive recursive predicate and hence definable in T as well as in HA, and since HA^ω is consistent, $con\ n$ is true and therefore derivable in T and in HA for any numeral n. The Π_1^0-statement

$$Con :\equiv \forall z\ con\ z$$

is a consistency statement for HA^ω which by Gödel's second incompleteness theorem is not derivable in HA^ω.

The sentence $D :\equiv Con \to \top \to Con$ is a tautology, derivable already in HA, and has the $\wedge t'$-translation

$$\exists X W \forall z (D \wedge (\forall x < Xz\ con(Wxz) \to ((\top \to Con) \wedge (\top \to con\ z))))$$

Are there functionals X and W that provably satisfy the matrix of this formula? Without really restricting the choice of X and W, we may require $Xz > 0$ and $W0z = z$. Under these conditions, HA^ω proves

$$(Con \to \top \to Con) \wedge (\forall x < Xz\ con(Wxz) \to (\top \to con\ z))$$

After cancellation of the now irrelevant $\top$, it remains to show for all z and suitable functionals X and W

$$\forall x < Xz\ con(Wxz) \to Con$$

This is impossible even for $z = 0$, because by normalization, $X0$ is provably equal to a numeral N, and for all $i < N$, the terms $Wi0$ are equal to numerals, whence all N formulae $con(Wi0)$ for $i < N$ and therefore also the antecedent $\forall x < X0 \; con(Wx0)$ are provable, while the consequent Con is not, as mentioned.

The translations $\wedge q$ and $\wedge t$ do not suffer of such a defect. Like mq and mrt, they are closely related:

1.7.3 *Proposition*

$$HA^\omega \vdash A_{\wedge t}[v, w] \leftrightarrow A \wedge A_{\wedge q}[v, w]$$

Proof by subformula induction. For atomic A, for conjunction and for bounded universal quantification, the claim is immediate or follows immediately from the I.H.

($\exists$) $(\exists u A[u])_{\wedge t} \equiv A_{\wedge t}[u, v, w]$
is by I.H. equivalent to $A[u] \wedge A_{\wedge q}[u, v, w]$ which is equivalent to

$$\exists u A[u] \wedge (A[u] \wedge A_{\wedge q}[u, v, w] \equiv \exists u A[u] \wedge (\exists u A[u])_{\wedge q}$$

($\forall$) $(\forall u A[u])_{\wedge t} \equiv \forall u A[u] \wedge A_{\wedge t}[u, Vu, w]$ is by I.H. equivalent to
$\forall u A[u] \wedge A[u] \wedge A_{\wedge q}[u, Vu, w]$ which is equivalent to $\forall u A[u] \wedge (\forall u A[u])_{\wedge q}$.
($\rightarrow$) $(A \rightarrow B)_{\wedge t} \equiv (A \rightarrow B) \wedge (A \wedge \forall x < X v z A_{\wedge t}[v, Wxvz] \rightarrow B_{\wedge t}[Yv, z])$
is by I.H. equivalent to
$(A \rightarrow B) \wedge (A \wedge \forall x < X v z (A \wedge A_{\wedge q}[v, Wxvz]) \rightarrow B \wedge B_{\wedge q}[Yv, z])$
which is equivalent to
$(A \rightarrow B) \wedge (A \wedge \forall x < X v z \, A_{\wedge q}[v, Wxvz] \rightarrow B_{\wedge q}[Yv, z])$
$\equiv (A \rightarrow B) \wedge (A \rightarrow B)_{\wedge q}$
$(\rightarrow)_0$ $(A \rightarrow B)_{\wedge t} \equiv (A \rightarrow B) \wedge (A_{\wedge t}[v] \rightarrow B_{\wedge t}[Yv, z])$
which is by I.H. equivalent to $(A \rightarrow B) \wedge (A \wedge A_{\wedge q}[v] \rightarrow B \wedge B_{\wedge q}[Yv, z])$
which is equivalent to
$(A \rightarrow B) \wedge (A \wedge A_{\wedge q}[v] \rightarrow B_{\wedge q}[Yv, z] \equiv (A \rightarrow B) \wedge (A \rightarrow B)_{\wedge q})$

By this proposition, the same term tuples $\wedge t$- and $\wedge q$-interpret the theorems of HA^ω. In fact, we shall show that there are such term tuples which in addition also $\wedge$-interpret these theorems:

1.7.4 *Uniform $\wedge -, \wedge q-,$ and $\wedge t-$ interpretation theorem for HA^ω*

Let $A^I \equiv \exists v \forall w A_I[v, w]$ for $I = \wedge, \wedge q, \wedge t$. If $HA^\omega \vdash A$, then there is a tuple b of terms from $L(T_\wedge)$ (of the same type tuple as v) for which at the same time

$$T_\wedge \vdash A_\wedge[b, w] \,, \quad HA^\omega \vdash A_{\wedge q}[b, w] \quad and \quad HA^\omega \vdash A_{\wedge t}[b, w]$$

So, besides $HA^\omega \overset{\wedge}{\hookrightarrow} T_\wedge$ (cf. theorem 1.2.7) , we have

$$HA^\omega \overset{\wedge q}{\hookrightarrow} HA^\omega \text{ and } HA^\omega \overset{\wedge t}{\hookrightarrow} HA^\omega$$

In view of proposition 1.7.3, it suffices to prove this theorem for $\wedge$ and $\wedge q$. In analogy to the procedure in the previous section, we prove that part in a more general form, this time subsuming $\wedge$ and $\wedge q$ under the translation schema $\wedge G$.

1.7.5 Recursive definition of functional translations $\wedge G$

Let G be an arbitrary sentence of $L(HA^\omega)$. To any formula A of HA^ω, $\wedge G$ assigns a formula $A^{\wedge G} \equiv \exists v \forall w A_{\wedge G}[v, w]$ of $L(HA^\omega)$ as follows:

$$(T_\wedge)^{\wedge G} \qquad A^{\wedge G} \equiv A \quad \text{for} \quad A \in L(T_\wedge)$$

Let $A^{\wedge G}$ be as above and $B^{\wedge G} \equiv \exists y \forall z B_{\wedge G}[y, z]$. Then

$$
\begin{aligned}
(\wedge)^{\wedge G} \quad &(A \wedge B)^{\wedge G} &\equiv\ & \exists v y \forall w z (A_{\wedge G} \wedge B_{\wedge G}) \\
(\rightarrow)^{\wedge G} \quad &(A \rightarrow B)^{\wedge G} &\equiv\ & \exists XWY \forall vz((G \rightarrow A) \wedge \\
& & & \forall x < X vz A_{\wedge G}[v, W x v z] \rightarrow B_{\wedge G}[Yv, z]) \\
& & & \text{in case the tuple } w \text{ is not empty} \\
(\rightarrow)_0^{\wedge G} \quad &(A \rightarrow B)^{\wedge G} &\equiv\ & \exists Y \forall vz((G \rightarrow A) \wedge A_{\wedge G}[v] \rightarrow B_{\wedge G}[Yv, z]) \\
& & & \text{for empty } w \\
(\forall <)^{\wedge G} \quad &(\forall x < t B[x])^{\wedge G} &\equiv\ & \exists Y \forall z\, \forall x < t B_{\wedge G}[Y x, z] \\
(\forall)^{\wedge G} \quad &(\forall u A[u])^{\wedge G} &\equiv\ & \exists V \forall u w A_{\wedge G}[u, V u, w] \\
(\exists)^{\wedge G} \quad &(\exists u A[u])^{\wedge G} &\equiv\ & \exists u v \forall w((G \rightarrow A[u]) \wedge A_{\wedge G}[u, v, w])
\end{aligned}
$$

Again, the cases where G is $\bot$ or G is $\top$ are relevant:
(i) $G \equiv \bot$. In this case, the conjunct $\bot \rightarrow A$ is true and may hence be omitted in the above definition. Thus, the $\wedge\bot$- translation becomes again the $\wedge$-translation.
(ii) $G \equiv \top$. Then, the conjunct $\top \rightarrow A$ is equivalent to A, and the $\wedge\top$-translation is in fact the $\wedge q$-translation.
It is an easy exercize to prove a lemma for the $\wedge G$-translation in analogy to lemma 1.2.12, in particular

$$HA^\omega \vdash (A \rightarrow B)^{\wedge G} \rightarrow ((G \rightarrow A) \wedge A^{\wedge G}) \rightarrow B^{\wedge G}$$

1.7.6 Uniform $\wedge G$-interpretation theorem for HA^ω

Let $A^{\wedge G} \equiv \exists v \forall w A_{\wedge G}[v, w]$. If $HA^\omega \vdash A$, then there is a tuple b of terms from $L(T_\wedge)$ for which for all sentences G

$$HA^\omega \vdash A_{\wedge G}[b, w]$$

Therefore, for any sentence G of $L(HA^\omega)$,

$$HA^\omega \overset{\wedge G}{\hookrightarrow} HA^\omega$$

The *proof* by induction on deductions parallels the proof of the $\wedge$-interpretation theorem 1.2.7. It turns out that the computation of the interpreting terms remains unchanged, i.e. it is independent of the conjuncts $G \to \dots$ that enter into the translated formulae. In order to amend the proof of 1.2.7, we only have to control the influence that the weakening conjuncts $G \to \dots$ have on the correctness of the translation of a given derivation. So, in the proof of 1.2.7, we substitute $\wedge G$ for $\wedge$ and abbreviate instances of $A_{\wedge G}, \dots$, and also of $\forall x < t\, A_{\wedge G}, \dots$, as they occur there, by $A', A'', \dots$.

Axioms $(K), (S), (P)$ (in the presence of product types), $(Axsuc)$, $(AxR0), (AxRsuc)$ are quantifier free, hence identical to their $\wedge G$-translation and interpreted trivially. Axioms $A1, A4, A\bot, B1$ to $B4$, and $Q1$ are implications $A \to B$ the $\wedge G$-translations of which have matrices of the form $(G \to A) \wedge A' \to B'$. Terms $\wedge$-interpreting $A \to B$, i.e. terms satisfying $(A \to B)_\wedge$, in these cases obviously also satisfy $A' \to B'$ and hence also $\wedge G$-interpret $A \to B$.

$A2$. The 2nd I.H. is weakened to

$$(G \to A) \wedge \forall x < Xvz A_{\wedge G}[v, Wxvz] \to B_{\wedge G}[Yv, z]$$

or, in short, to $(G \to A) \wedge A' \to B'$. However, since by the first premise, we have A and hence $G \to A$, we get $A' \to B'$ and can continue as in 1.2.7.

$A3$. The two I.H. are weakened to

$$(G \to A) \wedge A' \to B' \text{ and } (G \to B) \wedge B'' \to C'$$

respectively. However, since the 1st premise $A \to B$ by Frege's rule implies $(G \to A) \to (G \to B)$, we also get

$$(G \to A) \wedge B'' \to C'.$$

From this, by the argument in 1.2.7, we obtain $(G \to A) \wedge A'' \to C''$ with the same interpreting terms as before.

$A5/A6$. Since

$$(G \to A \wedge B) \leftrightarrow (G \to A) \wedge (G \to B),$$

is derivable, we have equivalence of derivability of

$$(G \to A \wedge B) \wedge A' \wedge B' \to C' \text{ and } (G \to A) \wedge A'' \to (G \to B) \wedge B'' \to C'$$

Hence, both rules are interpreted as in 1.2.7.

A7. By I.H. $(G \to A) \wedge A' \to B'$ and $(G \to A) \wedge A'' \to C'$, hence

$$(G \to A) \wedge A' \wedge A'' \to B' \wedge C'$$

This yields the interpretation of $A \to B \wedge C$ as in 1.2.7.

(Eq) The I.H. $(G \to A) \wedge A' \to a = b$ and $F'[a]$ leads by *(Eq)* to

$$(G \to A) \wedge A' \to F'[b]$$

which essentially is the $\wedge G$-interpretation of $A \to F[b]$.

(IND) From the first premise $A[0]$ and the first I.H. $A[0]'$, we obtain $(G \to A[0]) \wedge A[0]'$. The second premise $A[x] \to A[suc\, x]$ implies $(G \to A[x]) \to (G \to A[suc\, x])$. From this and the second I.H. $(G \to A[x]) \wedge A[x]' \to A[suc\, x]'$, we obtain

$$(G \to A[x]) \wedge A[x]' \to (G \to A[suc\, x]) \wedge A[suc\, x]'.$$

By the same calculation as in the proof of theorem 1.2.7, applying the generalized induction principle 1.1.22 (now as a theorem of HA^ω) to this longer induction formula, we get $(G \to A[t]) \wedge A[t]'$, hence $A[t]' \equiv A_{\wedge G}[t, ft, z]$ for the same terms f as in 1.2.7.

Q2. $(G \to A[b]) \wedge A[b]' \to (G \to A[u]) \wedge A[u]'$ holds for $u := b$ as in 1.2.7.

Q3. By I.H., $(G \to A) \wedge A' \to B[u]'$, with u not in G, A, A'. As in 1.2.7, by λ-abstraction, we get $(G \to A) \wedge A' \to (\forall u B[u])'$.

Q4. By I.H., $(G \to A[u]) \wedge A[u]' \to B'$. Again, by λ-abstraction, we get $(G \to \exists u A[u]) \wedge (G \to A[u]) \wedge A[u]' \to B'$ which is the desired instance of the $\wedge G$-matrix of $\exists u A \to B$.

This completes the amendation of the proof of the $\wedge$-interpretation theorem 1.2.7, filling in those parts which are needed to produce a proof of the uniform $\wedge G$-interpretation theorem 1.7.6.

Proof of theorem 1.7.4 By this last theorem, $\wedge \equiv \wedge\bot$ and $\wedge q \equiv \wedge\top$ are uniform functional interpretations of HA^ω. By proposition 1.7.3, this holds also for $\wedge q$ and $\wedge t$, as already mentionaed. So, theorem 1.7.4 is proved.

From this point onwards, we concentrate on the $\wedge q$-translation. The translation rules for $\wedge q$ differ in important points (4. and 6. below) from the rules for $\wedge$ as listed in lemma 1.2.12:

1.7.7 Lemma

In HA^ω is derivable:
1. $A^{\wedge q} \leftrightarrow A$ *for* $A \in L(T_\wedge)$;

2. $(A \wedge B)^{\wedge q} \leftrightarrow A^{\wedge q} \wedge B^{\wedge q}$;

3. $(\forall x < tB)^{\wedge q} \leftrightarrow \forall x < tB^{\wedge q}$;

4. $(\exists u A)^{\wedge q} \leftrightarrow \exists u (A \wedge A^{\wedge q})$;

5. $(\forall u A)^{\wedge q} \to \forall u A^{\wedge q}$;

6. $(A \to B)^{\wedge q} \to (A \wedge A^{\wedge q} \to B^{\wedge q})$; *furthermore*

7. $HA^\omega + AC \vdash \forall u A^{\wedge q} \to (\forall u A)^{\wedge q}$.

4., 5., *and* 7. *also hold for non-empty tuples u of variables.*

The *proof* is again left to the reader. This lemma holds in fact for $\wedge G$ in general, if, in 4. and 6., the untranslated A is replaced by $(G \to A)$. However, we do not know of any application of this more general result.

In spite of its formal proximity to the $\wedge$-translation, the $\wedge q$-translation is closer to mq and mr than to $\wedge$. We start to analyze the schema $\{A \leftrightarrow A^{\wedge q}\}$ by considering its direction $\{A \to A^{\wedge q}\}$ for Harrop formulae A. They do not have as simple a translation under $\wedge q$ as they have under mr and mq, but there is still a relation corresponding to lemma 1.6.14:

1.7.8 Proposition

For Harrop formulae A,

$$HA^\omega \vdash A \to A_{\wedge q}[v, w]$$

and therefore also

$$HA^\omega \vdash A \to A^{\wedge q}$$

Proof by subformula induction. In view of the previous lemma, it suffices to show the cases $(\forall)$ and $(\to)$, since $\exists$-formulae are not Harrop.

$(\forall)$ If by I.H. $A[u] \to A_{\wedge q}[u, v, w]$, then also $\forall u A[u] \to A_{\wedge q}[u, Vu, w]$.

$(\to)$ If by I.H. $B \to B_{\wedge q}[y, z]$ then as above $B \to B_{\wedge q}[Yv, z]$ and hence

$$(A \to B) \to (A \wedge \forall x < Xvz A_{\wedge q}[v, Wxvz] \to B_{\wedge q}[Yv, z])$$

which is $(A \to B) \to (A \to B)_{\wedge q}$.

From this we again obtain closure of HA^ω under the rules formulated in theorem 1.6.18 in the previous section:

1.7.9 Theorem

Also the $\wedge q$-interpretation shows closure of HA^ω under (Rule-IP$_H$) in its explicit form.

Proof. Let $HA^\omega \vdash A \to \exists y B[y]$ with A Harrop. By $\wedge q$-interpretation (cf. theorem 1.7.4), there are term tuples X, W, Y, Y_1 such that

$$HA^\omega \vdash A \wedge \forall x < Xvz\, A_{\wedge q}[v, Wxvz] \to B[Yv] \wedge B_{\wedge q}[Yv, Y_1v, z]$$

Since A is Harrop, $\vdash A \to \forall x < Xvz\, A_{\wedge q}[v, Wxvz]$ by the above proposition 1.7.8, and therefore

$$HA^\omega \vdash A \to B[Y0]$$

as was to be shown. Here, 0 is a tuple of 0-functionals of the same type tuple as v. We do not know whether there is any relation between the realizing terms c mentioned in theorem 1.6.18 and the $\wedge q$-interpreting terms $Y0$ here.

(ED) and *(Rule-AC)* now follow as in theorem 1.6.18.

By 1.7.8, the schema $\{A \to A^{\wedge q}\}$ does not add anything to HA^ω, if restricted to Harrop formulae A, but extended to the full language of HA^ω, it is at least as strong as the schemata $\{A \to A^{mq}\}$ and $\{A \to A^{mrt}\}$:

1.7.10   *Proposition*

In $HA^\omega + \{A \to A^{\wedge q}\}$, the schemata AC, IP_H, and $\{A^{\wedge q} \to A\}$ are provable.

Proof. 1. Let $C \equiv \forall x \exists y A[x, y]$. Then

$$C^{\wedge q} \equiv \exists Y V \forall x w(A[x, Yx] \wedge A_{\wedge q}[x, Yx, Vx, w])$$

So, $C^{\wedge q}$ implies $\exists Y \forall x A[x, Yx]$, and $C \to C^{\wedge q}$ implies AC.
2. Let $C \equiv A \to \exists y\, B[y]$ with A Harrop. Then by the argument in theorem 1.7.8,

$$C^{\wedge q} \to \exists Y (A \to B[Y0])$$

So, $C \to C^{\wedge q}$ implies IP_H.
3.   The schema $\{A^{\wedge q} \to A\}$ is proved by subformula induction on A. The cases for $A \in L(T_\wedge)$, conjunction, bounded and unbounded universal quantification are immediate.
($\exists$) holds by 1.7.7.4 even without I.H.
($\to$) By the axiom $A \to A^{\wedge q}$ and the I.H. $B^{\wedge q} \to B$, 1.7.7.6 implies

$$(A \to B)^{\wedge q} \to (A \to B)$$

This proposition, with mq or mrt in place of $\wedge q$, is an immediate consequence of the characterization theorem 1.6.25 and proposition 1.6.2. It implies that also the schemata $(IP_H)^{\wedge q}$ and $(AC)^{\wedge q}$ are provable in this theory. However, for IP_H we can do better:

1.7.11 Proposition

IP_H is $\wedge q$-interpretable in HA^ω. In particular,

$$HA^\omega \vdash (IP_H)^{\wedge q}$$

Proof. Let $C \equiv A \to \exists y B[y]$ with $C_{\wedge q}$ as explicitly stated in the proof of theorem 1.7.9. Since A is Harrop, we have by proposition 1.7.8

$$C_{\wedge q}[X, W, Y, Y_1; v', z'] \leftrightarrow (A \to B[Yv'] \wedge B_{\wedge q}[Yv', Y_1 v'; z']),$$

with the right hand side independent of X, W;
for $D \equiv \exists y (A \to B[y])$, we get similarly

$$D_{\wedge q}[X', W', y, Y_1'; v, z] \leftrightarrow (A \to B[y]) \wedge (A \to B_{\wedge q}[y, Y_1' v; z]),$$

with the right hand side independent of X', W'.
In order to $\wedge q$-interpret $C \to D$ in HA^ω, we have to find term tuples X_0, W_0, Y_0 satisfying

$$(C \to D)_{\wedge q} \equiv C \wedge \forall x < X_0 v_0 z_0 C_{\wedge q}[v_0, W_0 x v_0 z_0] \to D_{\wedge q}[Y_0 v_0, z_0]$$

Here, the free variables are $v_0 \equiv X, W, Y, Y_1$ and $z_0 \equiv v, z$, with X, W and also X', W' being irrelevant for the interpretation. We put

$$X_0 v_0 z_0 = 1 \text{ and } W_0 x v_0 z_0 = 0, z$$

For these X_0, W_0, $(C \to D)_{\wedge q}$ is implied by

$$C_{\wedge q}[v_0; 0, z] \to D_{\wedge q}[Y_0 v_0; v, z]$$

For $Y_0 v_0 = Y_0 X W Y Y_1 = X, W, Y0, \lambda v.(Y_1 0)$ (or $= 0, 0, Y0, \lambda v.(Y_1 0)$), this becomes

$$(A \to B[Y0] \wedge B_{\wedge q}[Y0, Y_1 0; z]) \to (A \to B[Y0]) \wedge (A \to B_{\wedge q}[Y0, \lambda v.(Y_1 0)v, z])$$

which is derivable in HA^ω.

This result does not imply that $HA^\omega + IP_H$ is $\wedge q$-interpretable in HA^ω. As with mq, the class of $\wedge q$-interpretable formulae is not deductively closed. We only have:

$$HA^\omega + IP_H \overset{\wedge q}{\hookrightarrow} HA^\omega + IP_H$$

The full schema AC is apparently not $\wedge q$-interpretable in $HA^\omega + IP_H + AC$, $(AC)^{\wedge q}$ seems to be underivable in this theory. We therefore first look for a class of formulae A for which the schema $\{A \to A^{\wedge q}\}$ does not exceed $HA^\omega + IP_H + AC$. Stein [39] observes that formulae without strictly negative occurrences of $\exists$, a *dual version* of Harrop formulae, will serve this purpose.

1.7.12 *Recursive definition of dual Harrop or dH formulae, and related AC- and IP-schemata*

1. The formulae of $L(T_\wedge)$ are dH.
2. The dH formulae are closed under conjunction, existential and (bounded and unbounded) universal quantification.
3. If A is Harrop and B is dH, then $A \to B$ is dH.

Moreover, we introduce the schemata

$AC_H \quad \forall x \exists y A[x, y] \to \exists Y \forall x A[x, Yx] \quad$ with A Harrop;
$AC_{dH} \quad \forall x \exists y A[x, y] \to \exists Y \forall x A[x, Yx] \quad$ with A dual Harrop;
$IP_{HdH} \quad (A \to \exists y B[y]) \to \exists y (A \to B[y])$ with A Harrop, B dual Harrop.

Axioms AC_{dH} as well as IP_{HdH} are both implications between two dual Harrop formulae. By restricting the proof of proposition 1.7.10 to dual Harrop formulae, we obtain:

1.7.13 *Corollary to proposition 1.7.10*

In $HA^\omega + \{A \to A^{\wedge q} \mid A \text{ dual Harrop}\}$, the schemata AC_{dH} and IP_{HdH} are derivable.

1.7.14 *Lemma*

If A is dual Harrop,

$$HA^\omega \vdash A^{\wedge q} \to A$$

The *proof* by subformula induction proceeds by 1,2,3,4,5 of lemma 1.7.7. In the case of implications $A \to B$, A is Harrop and B is dual Harrop, so that HA^ω proves $A \to A^{\wedge q}$ by proposition 1.7.8 and $B^{\wedge q} \to B$ by I.H. This, together with lemma 1.7.7.6, implies

$$HA^\omega \vdash (A \to B)^{\wedge q} \to (A \to B)$$

With respect to dual Harrop formulae, $\wedge q$ behaves like mrt with respect to the full language.

1.7.15 *Partial characterization theorem for $\wedge q$*

The following theories are equivalent:

$$HA^\omega + \{A \leftrightarrow A^{\wedge q} \mid A \text{ dual Harrop }\}$$
$$HA^\omega + AC_{dH} + IP_{HdH}$$
$$HA^\omega + AC_{\exists free} + IP_{\exists free}$$
$$HA^\omega + \{A \leftrightarrow A^{mr}\}$$
$$HA^\omega + AC + IP_H$$

Proof. By corollary 1.7.13, $HA^\omega + \{A \leftrightarrow A^{\wedge q} \mid A$ dual Harrop $\}$ proves the schemata $AC_{dH} + IP_{HdH}$ which contain the schemata $AC_{\exists free} + IP_{\exists free}$, because $\exists$free formulae are Harrop as well as dual Harrop. These last schemata are, by the characterization theorem for mr 1.6.17, equivalent to $\{A \leftrightarrow A^{mr}\}$ as well as to $AC + IP_H$ which in turn contain $AC_{dH} + IP_{HdH}$. Because of lemma 1.7.14, it remains to show for dual Harrop formulae A

$$HA^\omega + AC + IP_H \vdash A \to A^{\wedge q}$$

This is done by subformula induction. We discuss the cases:

($\forall$) The I.H. together with lemma 1.7.7.7 implies $\forall u A[u] \to (\forall u A[u])^{\wedge q}$.

($\exists$) The I.H. together with lemma 1.7.7.4 implies $\exists u A[u] \to (\exists u A[u])^{\wedge q}$

($\to$) By I.H., since B is dual Harrop, $A \to B$ implies $A \to B^{\wedge q}$ which by IP_H implies $\exists y (A \to \forall z B_{\wedge q}[y, z])$, because A is Harrop. This logically implies $(A \to B)^{\wedge q}$.

It remains to close the gap between this partial and a full characterization theorem.

1.7.16 *Lemma*

Let Γ be a class of formulae from which the schema AC is provable. Then

$$HA^\omega + \Gamma \vdash A \text{ implies } HA^\omega + \Gamma + \Gamma^{\wedge q} \vdash A^{\wedge q}$$

Proof. By the deduction theorem, applied to the premise, there are $C_1, ..., C_n \in \Gamma$ such that

$$HA^\omega \vdash \forall C_1 \wedge ... \wedge \forall C_n \to A$$

By the $\wedge q$-interpretation theorem 1.7.4,

$$HA^\omega \vdash (\forall C_1 \wedge ... \wedge \forall C_n \to A)^{\wedge q}$$

This implies by lemma 1.7.7.6 and 1.7.7.2

$$HA^\omega \vdash \forall C_1 \wedge ... \wedge \forall C_n \wedge (\forall C_1)^{\wedge q} \wedge ... \wedge (\forall C_n)^{\wedge q} \to A^{\wedge q}$$

and therefore

$$HA^\omega + \Gamma \vdash (\forall C_1)^{\wedge q} \wedge ... \wedge (\forall C_n)^{\wedge q} \to A^{\wedge q}.$$

Since $HA^\omega + \Gamma$ proves AC, we may by lemma 1.7.7.7 replace $(\forall C_i)^{\wedge q}$ by $\forall (C_i^{\wedge q})$ for $i = 1, ..., n$ and obtain

$$HA^\omega + \Gamma + \Gamma^{\wedge q} \vdash A^{\wedge q}.$$

1.7.17   *Axiomatization theorem for* $\{A \to A^{\wedge q}\}$

For all formulae A,

$$HA^\omega + AC_{\exists free} + IP_{\exists free} + (AC_{\exists free})^{\wedge q} \vdash A \to A^{\wedge q}$$

Proof. By propositon 1.6.12,

(1) $HA^\omega + AC_{\exists free} + IP_{\exists free} \vdash A^{mr} \to A$

By the characterization theorem for mr 1.6.17, $HA^\omega + AC_{\exists free} + IP_{\exists free}$ proves the schema AC. Therefore, lemma 1.7.16 is applicable, and since $HA^\omega \vdash (IP_{\exists free})^{\wedge q}$ by proposition 1.7.11,

(2) $HA^\omega + AC_{\exists free} + IP_{\exists free} + (AC_{\exists free})^{\wedge q} \vdash (A^{mr} \to A)^{\wedge q}$

By lemma 1.7.7.6,

(3) $HA^\omega \vdash (A^{mr} \to A)^{\wedge q} \to (A^{mr} \wedge (A^{mr})^{\wedge q} \to A^{\wedge q})$

But v mr A is $\exists$free and therefore Harrop, so that by proposition 1.7.8 $HA^\omega \vdash v$ mr $A \to (v$ mr $A \wedge (v$ mr $A)^{\wedge q})$ and therefore by 1.7.7.4 also

$$HA^\omega \vdash A^{mr} \to (A^{mr})^{\wedge q}$$

So we may cancel $(A^{mr})^{\wedge q}$ in (3) and obtain

(4) $HA^\omega \vdash (A^{mr} \to A)^{\wedge q} \to (A^{mr} \to A^{\wedge q})$

and by (2)

(5) $HA^\omega + AC_{\exists free} + IP_{\exists free} + (AC_{\exists free})^{\wedge q} \vdash A^{mr} \to A^{\wedge q}$

In (1), we used only one direction of the equivalence $A \leftrightarrow A^{mr}$ of proposition 1.6.12; now we apply the reverse direction $A \to A^{mr}$ to (5) and thus finish the proof. - As in the proof of proposition 1.6.23, the $\exists$-*prenex normal form* A^{mr} of A is used here to mediate between A and the hybrid translation $A^{\wedge q}$ resp. A^{mrG} of A.

This axiomatization theorem, together with proposition 1.7.10, allows a characterization of the schema $\{A \to A^{\wedge q}\}$ in terms of $(AC_{\exists free})^{\wedge q}$ which still depends on the translation $\wedge q$. Such a characterization is unsatisfactory. Stein [39] $\wedge q$-translates $AC_{\exists free}$, rather the more general schema AC_H (cf. definition 1.7.12) into a schema that does not mention $\wedge q$.

1.7.18   *Stein's principle of independent choice*

IC_H $\forall Y \exists ZX (\forall z < ZA[Xz, Y(Xz)] \to \exists Y_0 (\forall x \exists y A[x, y] \to \forall x A[x, Y_0 x]))$

for Harrop formulae A.

This principle may be read as follows:

Let $A[x^\sigma, y^\tau]$ be a Harrop formula. For any function $Y : \sigma \to \tau$, there is a "critical" finite set $\{Xz \mid z < Z\}$ with the following property: If Y is a choice function for $A[x, y]$ on that critical finite set, then a function $Y_0 : \sigma \to \tau$ can be found which is a choice function for all of A, provided that $\forall x \exists y A[x, y]$.

1.7.19   Proposition

On the basis of $HA^\omega + AC + IP_H$, the schema $(AC_H)^{\wedge q}$ is equivalent to Stein's principle of independent choice IC_H.

Proof. Let $C \equiv \forall x \exists y A[x, y]$ with A Harrop. Then

$$C^{\wedge q} \equiv \exists Y V \forall x w (A[x, Yx] \wedge A_{\wedge q}[x, Yx, Vx, w])$$

which is equivalent to $\exists Y \forall x A[x, Yx]$ by proposition 1.7.8.
Let $D \equiv \exists Y_0 \forall x A[x, Y_0 x]$. Then

$$D^{\wedge q} \equiv \exists Y_0 V \forall x w ((\forall x A[x, Y_0 x]) \wedge A_{\wedge q}[x, Y_0 x, Vx, w])$$

is equivalent to $\exists Y_0 (\forall x A[x, Y_0 x])$, again by 1.7.8. That is in fact the same formula as the short version of $C^{\wedge q}$, but with a different division into prefix and matrix. In $D^{\wedge q}$, the formula $\forall x A[x, Y_0 x]$ is part of and equivalent to the matrix $D_{\wedge q}$. Taking that into account, it suffices to use these short versions of $C^{\wedge q}$ and $D^{\wedge q}$ for the translation of AC_H. Then $(AC_H)^{\wedge q} \equiv (C \to D)^{\wedge q}$ is equivalent to

$$\exists Z X Y_0 \forall Y (\forall x \exists y A[x, y] \wedge \forall z < ZY A[XzY, Y(XzY)] \to \forall x A[x, Y_0 Yx])$$

which is now independent of the $\wedge q$-translation of A. In this formula, $\forall z < ZY A[XzY, Y(XzY)]$ is Harrop, whereas $\forall x \exists y A[x, y]$ is not. We therefore rearrange equivalently to

$$\exists Z X Y_0 \forall Y (\forall z < ZY A[XzY, Y(XzY)] \to \forall x \exists y A[x, y] \to \forall x A[x, Y_0 Yx])$$

which implies IC_H directly and is implied by IC_H via IP_H and AC.

Collecting these results, we have:

1.7.20   Characterization theorem for $\wedge q$, Stein [39]

The following theories are equivalent:

$$HA^\omega + \{A \leftrightarrow A^{\wedge q}\}$$
$$HA^\omega + \{A \to A^{\wedge q}\}$$
$$HA^\omega + AC_{\exists free} + IP_{\exists free} + (AC_{\exists free})^{\wedge q}$$
$$HA^\omega + AC + IP_H + (AC)^{\wedge q}$$
$$HA^\omega + AC + IP_H + (AC_H)^{\wedge q}$$
$$HA^\omega + + AC + IP_H + IC_H$$

Proof. By proposition 1.7.10, the first two theories are equivalent and prove the axiom schemata of the next three theories. By the axiomatization theorem 1.7.17, already the third theory – which is extended by the next two theories – proves the schema $\{A \to A^{\wedge q}\}$. The last two theories are equivalent by proposition 1.7.19.

Stein's characterization of $\wedge q$ transfers also to $\wedge t$:

1.7.21 Characterization theorem for $\wedge t$

On the basis of HA^{ω}, the schemata $\{A \leftrightarrow A^{\wedge t}\}$ and $\{A \leftrightarrow A^{\wedge q}\}$ are equivalent: The functional interpretations $\wedge t$ and $\wedge q$ are characterized by the same schemata listed in the characterization theorem for $\wedge q$ above.

Proof. Since $\vdash A^{\wedge t} \to A$ by proposition 1.7.3, $A \leftrightarrow A^{\wedge t}$ is equivalent to $A \to A^{\wedge t}$, hence again by proposition 1.7.3 to $A \to A \wedge A^{\wedge q}$. i.e. to $A \to A^{\wedge q}$. However, by proposition 1.7.10, the schema $\{A \to A^{\wedge q}\}$ is equivalent to the schema $\{A \leftrightarrow A^{\wedge q}\}$. So the characterization theorem also for $\wedge t$ is proved.

1.8   Between $\wedge$-interpretation and modified realization

In this section, we present a unified treatment of $\wedge$-interpretation and modified realization together with an infinity of n-interpretations inbetween these two. This work is essentially a syntactic version of [11] and is related to work by Stein [40] and Rath [32].

The essential difference between the translations $\wedge$ (cf. 1.2.5) and mr (cf.1.6.1) is in the translation of implication. In the simplest case, for $A[w^{\tau}]$ and B essentially quantifier free, we may write simultaneously for $J = \wedge$ as well as for $J = mr$

$$(\forall w^{\tau} A[w^{\tau}] \to B)^{J} \equiv \exists\, set\, W^{\bar{\tau}}(\forall w^{\tau} \in W^{\bar{\tau}} A[w^{\tau}] \to B)$$

For $J = mr$, $W^{\bar{\tau}}$ is simply the set of all objects of type τ; for $J = \wedge$, $W^{\bar{\tau}}$ is a finite set of objects of type τ, enumerated by a function $W : o \to \tau$ up to a number $X : o$. For a common, general formulation we introduce an extension T_{J} of T which explicitly employs a restricted quantifier $\forall x^{\tau} \in a$ for all types τ (instead of the bounded universal quantifier $\forall x < t$) .

1.8.1 Description of the theory T_J as an extension of T in linear types

1. **J-types** are defined inductively by (1) and (2) in 0.1.1 and
(3) If τ is a J-type, then $\bar{\tau}$ is a J-type.
J-types of the form $\bar{\tau}$ are also called **set types**.

2. **Constants** of T_J are the constants of T (cf. 1.1.1, transferred to J-types) plus the following for all J-types τ, σ:
$$\{\,\} : \tau \to \bar{\tau}$$
$$\cup : \bar{\tau} \to \bar{\tau} \to \bar{\tau}$$
$$\bigcup : (\sigma \to \bar{\tau}) \to \bar{\sigma} \to \bar{\tau}$$

3. **J-terms** are defined inductively as in 0.1.3.1 and in 1.1.1, based on the constants of T_J (in J-types, hence without pairing functionals). However, we write
$\{a\}$ for $(\{\,\}a)$
$(a \cup b)$ for $((\cup a)\, b)$
$\bigcup\{b[x^\sigma] \mid x^\sigma \in a\}$ for $((\bigcup(\lambda x^\sigma.b[x^\sigma]))a)$

4. **Formulae of T_J** are defined inductively as follows:
4.1. If a, b are J-terms of the same linear type τ, then the equation $a = b$ of type τ is a formula of T_J. $\bot$ is a formula of T_J.
4.2. $L(T_J)$ is - like $L(T)$ - closed under conjunction $\wedge$ and implication $\to$.
4.3. (Restricted quantification) For any J-type τ, if a is a J-term of J-type $\bar{\tau}$ not containing the variable x^τ, and B is a formula, then $\forall x^\tau \in a\, B$ is a formula of T_J.

5. **Axioms and rules of T_J** are the axioms and rules of T (in the language of T_J) plus the following laws for the restricted quantifier:

$$J1 \quad \forall x \in \{a\}\, B[x] \qquad\qquad \to B[a]$$
$$J2 \quad \forall x \in (a \cup b)\, C \qquad \to \forall x \in a\, C$$
$$ \forall x \in (a \cup b)\, C \qquad \to \forall x \in b\, C$$
$$J3 \quad \forall y \in \bigcup\{b[x] \mid x \in a\}\, C \to \forall x \in a\, \forall y \in b[x]\, C$$
$$ (x \text{ not free in } C)$$
$$J4 \quad \forall x \in a(B[x] \to C[x]) \quad \to \forall y \in a\, B[y] \to \forall z \in a\, C[z]$$
$$J5 \quad A \to B \qquad\qquad\qquad \vdash A \to \forall x \in a\, B \quad (x \text{ not free in } a, A)$$

That completes the definition of T_J.

Results shown for the theories T in sections 1.1 and 1.4 obviously extend to T_J. Arguments concerning the bounded quantifier $\forall x < t$ in section 1.1 may often be transferred to the restricted quantifier $\forall y \in a$. In particular, we have in T_J:

1.8.2 *Lemma*

1. $B \vdash \forall x \in a\, B$
2. $\forall y \in a\, B[y] \to \forall z \in a\, B[z]$ (y, z new variables)
3. $A \to B \to C \vdash A \to \forall x \in a\, B \to \forall x \in a\, C$ (x not free in a)
4. $A \to B \to C \vdash \forall x \in a\, A \to \forall x \in a\, B \to \forall x \in a\, C$
5. $\forall x \in a\, A \wedge \forall x \in a\, B \leftrightarrow \forall x \in a\, (A \wedge B)$
6. $\forall x \in a\, A \wedge \forall y \in b\, B \to \forall x \in a\, \forall y \in b\, (A \wedge B)$
   (x not free in B, b, and y not free in A).

The *proofs* are left to the reader.

1.8.3 *Recursive definition of* 0-*functionals* 0^τ *for J-types* τ

1. $0^o := 0$
2. $0^{\sigma \to \tau} := \mathsf{K}_{\sigma\tau} 0^\tau$
3. $0^{\bar\tau} := \{0^\tau\}$

We have $\forall x \in 0^{\bar\tau} A[x] \to A[0^\tau]$ by $J1$.

A converse of lemma 1.8.2.6 holds only in the following form:

1.8.4 *Lemma*

$$T_J \vdash \forall x \in a \cup 0\, \forall y \in b \cup 0\, (A \wedge B) \to \forall x \in a\, A \wedge \forall y \in b\, B$$
(x not free in B, b, and y not free in A).

The *proof* applying $J1$, $J2$, and 3. and 5. of lemma 1.8.2 is again left to the reader.

1.8.5 *Proposition: Extended rule of induction for* T_J

The following extended rule of induction is admissible in T_J:

$$B[0, z],\ \forall w \in a\, B[x^o, w] \to B[suc\, x^o, z] \vdash B[t, z],$$

where z, w are disjoint tuples of variables of type tuple τ, a is a term tuple of type tuple $\bar\tau$, and t is substitutable in B.

Proof. Let $W := \lambda x^o z.a$ and define by simultaneous recursion, theorem 1.4.5

$$g0tz \quad\ = \{z\}$$
$$g(suc\, x)tz = \bigcup \{W(t \overset{.}{-} suc\, x)u \mid u \in gxtz\}$$

From the first assumption, we obtain by lemma 1.8.2.1

(1) $\forall u \in gttz\, B[0, u]$

By $J3$ and lemma 1.8.2.2, the definition of g yields

$$\forall u \in g(suc\,x)tz\, B[t \dot- suc\,x, u]$$
$$\to \forall u \in gxtz \forall w \in W(t \dot- suc\,x)u\, B[t \dot- suc\,x, w]$$

From this and by a substitution instance of the second assumption, we get by lemma 1.8.2.3 and $A3$

$$\forall u \in g(suc\,x)tz\, B[t \dot- suc\,x, u] \to \forall u \in gxtz\, B[suc(t \dot- suc\,x), u]$$

Since $x < t \to suc(t \dot- suc\,x) = t \dot- x)$, this implies

(2) $x < t \to \forall u \in g(suc\,x)tz\, B[t \dot- suc\,x, u] \to \forall u \in gxtz\, B[t \dot- x), u]$

By an induction "from $x = t$ down to $x = 0$" (cf. the proof of proposition 1.1.22), (1) and (2) imply

$$\forall u \in g0tz\, B[t \dot- 0, u]$$

which is $\forall u \in \{z\}\, B[t, u]$ which implies $B[t, z]$ by $J1$.

The argument for this proposition is conceptually similar to the proof of proposition 1.1.22.

We now come to the functional interpretation J of HA^ω for which the theory T_J has been tailored.

To avoid confusion with the restricted quantifier $\forall w \in a$, in this section, $\forall x < t$ is considered a defined expression, defined by

$$\forall x < t\, A[x] \equiv \forall x(x < t \to A[x])$$

The translation J is formulated in close analogy to the translation $\wedge$ in 1.2.5.

1.8.6 *Recursive definition of the functional translation J of $L(HA^\omega)$*

To any formula A of $L(HA^\omega)$, J assigns an expression $A^J \equiv \exists v \forall w\, A_J[v, w]$ with $A_J[v, w]$ a formula of $L(T_J)$ and tuples v, w of J-variables analogous to definition 0.1.15.

$(T)^J$ $A^J \equiv A$ for $A \in L(T)$

Let A^J be as above and $B^J \equiv \exists y \forall z\, B_J[y, z]$, then

$(\wedge)^J$ $(A \wedge B)^J \equiv \exists v y \forall w z (A_J[v, w] \wedge B_J[y, z])$

$(\to)^J$ $(A \to B)^J \equiv \exists WY \forall vz(\forall w \in Wvz\, A_J[v, w] \to B_J[Yv, z])$

$(\forall)^J$ $(\forall u A[u])^J \equiv \exists V \forall uw A_J[u, Vu, w]$

$(\exists)^J$ $(\exists u A[u])^J \equiv \exists uv \forall w A_J[u, v, w]$

In $(\to)^J$, the tuple W has the same length as the tuple w. If w is empty, i.e. if $A^J \equiv \exists v A_J[v]$, then

$$(A \to B)^J \equiv \exists Y \forall v z \, (A_J[v] \to B_J[Yv, z]).$$

So, translations J and $\wedge$ agree also in this point and therefore also for disjunctions:

$(\vee)^J \qquad (A \vee B)^J \equiv \exists xvy \forall wz((x = 0 \to A_J) \wedge (x \neq 0 \to B_J))$

The statement of lemma 1.2.6 obviously transfers to the J-translation:

1.8.7   *Lemma*

If $A^J \equiv B^J$, then $A \to B$ is J-interpretable in T_J.

Proof. By assumption, $A_J[v, z] \equiv B_J[v, z]$, so that by $J1$, $\forall x \in \{v\} A_J[x, z] \to B_J[v, z]$. So, $\lambda vz.\{v\}, \lambda v.v$ J-interpret $A \to B$.

1.8.8   *J-interpretation theorem for HA^ω*

The theory HA^ω is J-interpretable in T_J:

$$HA^\omega \overset{J}{\hookrightarrow} T_J$$

The *proof* of this theorem parallels the proof of the $\wedge$-interpretation theorem 1.2.7, but it is technically simpler, due to the smaller number of existential quantifiers in $(\to)^J$.

$A1$ is an instance of lemma 1.8.7.

$A2$. Given v_0 and W, Y such that by I.H. $A_J[v_0, w]$ and

$$\forall w \in W vz \, A_J[v, w] \to B_J[Yv, z],$$

lemma 1.8.2.1 implies $\forall w \in W v_0 z \, A_J[v_0, w]$, and after substitution of v_0 for v in the second I.H. we obtain $B_J[Y v_0, z]$ by $A2$.

$A3$. We may assume that all variables free in B also occur free in $A \to C$. By I.H., there are Y_1, W_1, Y_2, W_2 such that

$$\forall w \in W_1 vz \, A_J[v, w] \to B_J[Y_1 v, z] \text{ and } \forall w \in W_2 yx \, B_J[y, z] \to C_J[Y_2 y, x]$$

Substituting $Y_1 v$ for y in the right formula, applying lemma 1.8.2.3 to the right formula, and writing a for $W_2(Y_1 v)x$, we obtain

$$\forall z \in a \, \forall w \in W_1 vz \, A_J[v, w] \to \forall z \in a \, B_J[Y_1 v, z] \text{ and}$$
$$\forall z \in a \, B_J[Y_1 v, z] \to C_J[Y_2(Y_1 v), x]$$

By $J3$ and $A3$, this implies
$$\forall w \in \bigcup \{W_1 v z \mid z \in a\} \, A_J[v, w] \to C_J[Y_2(Y_1 v), x]$$
So W, Y defined by $W v x = \bigcup \{W_1 v z \mid z \in a\}$ and $Y v = Y_2(Y_1 v)$ J-interpret
$A \to C$.

$A4$ is "almost" an instance of lemma 1.8.7.

$A5$. By I.H., there are Y, W, W' such that
$$\forall w \in W v v' z \, \forall w' \in W' v v' z \, (A_J[v, w] \wedge B_J[v', w']) \to C_J[Y v v', z]$$
By lemma 1.8.2.6, Y, W, W' at the same time J-interpret $A \to B \to C$.

$A6$. By I.H., there are Y, W, W' such that
$$\forall w \in W v v' z \, A_J[v, w] \to \forall w' \in W' v v' z \, B_J[v', w'] \to C_J[Y v v', z]$$
By lemma 1.8.4, this implies
$$\forall w \in W v v' z \cup 0 \, \forall w' \in W' v v' z \cup 0 \, (A_J[v, w] \wedge B_J[v', w']) \to C_J[Y v v', z]$$
So Y and W_0, W_0' defined by $W_0 v v' z = W v v' z \cup 0$ and $W_0' v v' z = W' v v' z \cup 0$
J-interpret $A \wedge B \to C$.

$A7$. By I.H., there are Y, W and Y', W' such that
$$\forall w \in W v z \, A_J[v, w] \to B_J[Y v, z] \text{ and } \forall w \in W' v z' \, A_J[v, w] \to B_J[Y' v, z']$$
Thus, by $J2$, $A3$, and $A7$,
$$\forall w \in (W v z \cup W' v z') \, A_J[v, w] \to (B_J[Y v, z] \wedge C_J[Y' v, z'])$$
Hence, Y, Y', Z with $Z v z z' = W v z \cup W' v z'$ J-interpret $A \to B \wedge C$.

$A \bot_J$ is again an axiom $A \bot$.

Equality axioms $(K), (S), (R)$ and $(Ax\ suc)$ are their own J-translation and
J-interpreted trivially.

(Eq). By I.H., there are W', Y' with
$$\forall w \in W' v \, A_J[v, w] \to a = b \text{ and } F_J[a, Y', z]$$
This implies $\forall w \in W' v \, A_J[v, w] \to F_J[b, Y', z]$. So W, Y with $W v z = W' v$
and $Y v = Y'$ J-interpret the conclusion $A \to F[b]$.

(IND). By I.H., there are v_0, Y and W such that
$$F_J[0, v_0, z] \text{ and } \forall w \in W x v z \, F_J[x, v, w] \to F_J[suc\, x, Y x v, z]$$
By simultaneous recursion 1.4.5, we define a tuple f by
$$f0 = v_0 \text{ and } f(suc\, x) = Y x (f x)$$
Substituting $f x$ for v, rule (Eq) yields
$$F_J[0, f0, z] \text{ and } \forall w \in W x (f x) z \, F_J[x, f x, w] \to F_J[suc\, x, f(suc\, x), z]$$
By proposition 1.8.5 with $B[x, z] \equiv F_J[x, f x, z]$ and $a \equiv W x (f x) z$, this
implies $F_J[t, f t, z]$, and the tuple $f t$ J-interprets $F[t]$.

The J-interpretation of the quantifier laws $Q1$ to $Q4$ is similar to their
$\wedge$-interpretation in 1.2.7 and is left to the reader.

For a characterization of the translation J, we briefly look at extensions
HA_J^ω and $HA_J^{\omega+}$ of HA^ω as well as of T_J.

1.8.9 Definition of extensions HA_J^ω and $HA_J^{\omega+}$ of HA^ω and of T_J, together with the trivial extension of the functional interpretation J to these theories

The language of HA_J^ω is the extension of $L(T_J)$ by unbounded quantifiers $\exists x^\sigma, \forall y^\tau$ occurring, however, only outside the scope of restricted quantifiers $\forall z^\rho \in a$.

Axioms and rules of HA_J^ω are those of T_J extended to this larger language, plus the quantifier laws $Q1$ to $Q4$. So, HA_J^ω is also an extension of HA^ω. $HA_J^{\omega+}$ is, in analogy to $HA^{\omega+}$, the extension of HA_J^ω by the schemata

$$IP_J \quad (\forall w A \to \exists y B[y]) \to \exists y (\forall w A \to B[y]) \qquad \text{for } A \in L(T_J)$$
$$M_J \quad (\forall w A[w] \to B) \to \exists W (\forall w \in W\; A[w] \to B) \text{ for } A, B \in L(T_J)$$
$$AC_{\sigma,\tau} \; \forall x^\sigma \exists y^\tau A[x,y] \to \exists Y^{\sigma\to\tau} \forall x\, A[x, Yx]$$

w, y stand again for non-empty tuples of variables of arbitrary type-tuple. AC_J designates the union of all schemata $AC_{\sigma,\tau}$ (in $L(HA_J^\omega)$).

$$HA_J^{\omega+} :\equiv HA_J^\omega + IP_J + M_J + AC_J$$

Finally, for A in $L(T_J)$, we put $A^J \equiv A_J \equiv A$.

So, the J-interpretation of HA^ω extends immediately to HA_J^ω and, since J is idempotent, also to $HA_J^\omega + \{A \leftrightarrow A^J\}$:

1.8.10 Corollaries to the J-interpretation theorem 1.8.8

(1) $HA_J^\omega \overset{J}{\hookrightarrow} T_J$

(2) $HA_J^\omega + \{A \leftrightarrow A^J\} \overset{J}{\hookrightarrow} T_J$,

(3) HA_J^ω and $HA_J^\omega + \{A \leftrightarrow A^J\}$ are conservative extensions of T_J.

(4) $HA_J^\omega + \{A \leftrightarrow A^J\}$ is J-maximal relative to T_J

Also the characterization of the J-translation is analogous to the characterization of $\wedge$ in theorem 1.2.13:

1.8.11 Lemma

In HA_J^ω is derivable:
1. $A^J \leftrightarrow A$ *for* $A \in L(T_J)$;
2. $(A \wedge B)^J \leftrightarrow A^J \wedge B^J$;
3. $(\exists u A)^J \leftrightarrow \exists u A^J$;
4. $(\forall u A)^J \to \forall u A^J$;
5. $(A \to B)^J \to A^J \to B^J$; *furthermore*

6. $HA_J^\omega + AC \vdash \forall u A^J \to (\forall u A)^J$;

7. $HA_J^{\omega+} \vdash (A^J \to B^J) \to (A \to B)^J$.

The *proof* (for 5. and 7.) may be taken over from lemma 1.2.12. The following is a direct consequence of this lemma, the J-interpretation theorem 1.8.8, and its corollaries 1.8.10 :

1.8.12 *Characterization and extended interpretation theorem for the J-translation*

(1) *The theories $HA_J^{\omega+}$ and $HA_J^\omega + \{A \leftrightarrow A^J\}$ are equivalent:*

$$HA_J^{\omega+} \equiv HA_J^\omega + \{A \leftrightarrow A^J\}$$

(2) $HA_J^{\omega+}$ *is J-interpretable in T_J:*

$$HA_J^{\omega+} \overset{J}{\hookrightarrow} T_J$$

(3) $HA_J^{\omega+}$ *is J-maximal relative to T_J.*

An immediate consequence of the J-interpretability of Markov's schema M_J is:

1.8.13 *Corollary*

HA_J^ω *is closed under a form of Markov's rule:*
$(Rule - M_J)$ *If* $\vdash \forall w A \to B$ *with* $A, B \in L(T_J)$,
  then there are terms W s.t. $\vdash \forall w \in W \, A \to B$.

The theorem, as it stands, implies closure of HA_J^ω under $(Rule - AC)$ and $(Rule - IP_J)$ only for formulae $A, B \in L(T_J)$. In a different context, Stein [40] studies a q-hybrid of J to obtain such closure properties in general. This, however, requires an extension of HA_J^ω by allowing unbounded quantification within the scope of the restricted quantifier and also further axioms for T_J to allow for a J-interpretation of this extension of HA_J^ω in itself.

The restricted quantifier $\forall x \in a$ was introduced in T_J to make possible a unified treatment of different functional interpretations of HA^ω, depending on how this restricted quantifier is translated or *read* in the language of HA^ω. We introduce such *readings* ϕ of the theory T_J in HA^ω so that the J-interpretation of HA^ω in T_J is transformed via ϕ into functional interpretations $J\phi$ of HA^ω in HA^ω:

$$HA^\omega \overset{J}{\hookrightarrow} T_J \overset{\phi}{\longrightarrow} HA^\omega \quad \text{leads to} \quad HA^\omega \overset{J\phi}{\hookrightarrow} HA^\omega$$

The central point is the reading (or translation) of the restricted quantifier $\forall x^\sigma \in a$ which, for some or all types σ, may be ϕ-translated into the unrestricted quantifier $\forall x^{\phi\sigma}$, independent of $a : \bar{\sigma}$. In that case, $\phi\bar{\sigma}$ is an "improper" type, here called ω, of which the universe of objects of type $\phi\sigma$ may be taken to be the only element, here called 0^ω.

1.8.14   *Definition of readings ϕ of T_J in HA^ω*

A **reading** ϕ translates types, terms, and formulae of T_J as follows:
(1) $\phi : \{J - types\} \to \{linear\ types\} \cup \{\omega\}$ by
(1.1) $\phi(o) = o$
(1.2) $\phi(\sigma \to \tau) = \phi\sigma \to \phi\tau$, if $\phi\sigma$ and $\phi\tau$ are both linear
(1.3) $\phi(\sigma \to \tau) = \phi\tau$, if $\phi\sigma = \omega$ or $\phi\tau = \omega$
So $\phi\tau = \tau$ for all linear types τ.

(2) In *reading* the terms of T_J in HA^ω by ϕ,

$$\phi : \{J - terms\} \to \{terms\ of\ T\} \cup \{0^\omega\}$$

we take care that typing and reading commute, i.e. that for all J-terms a and J-types τ

$$a : \tau \quad \text{implies} \quad \phi a : \phi\tau$$

in particular $\phi a \equiv a$, if $a \in L(T_\wedge)$ and $\phi a \equiv 0^\omega$, if $a : \tau$ with $\phi\tau = \omega$.
For application, this means for $a : \sigma \to \tau$ and $b : \sigma$
(2.1) $\phi(ab) \equiv (\phi a)(\phi b) : \phi\tau$ if $\phi\sigma, \phi\tau$ are both linear
(2.2) $\phi(ab) \equiv \phi a$, if $\phi\tau$ is linear, $\phi\sigma = \omega$
(2.3) $\phi(ab) \equiv 0^\omega$, if $\phi\tau = \omega$
For variables x^τ, it means
(2.4) $\phi x^\tau \equiv x^{\phi\tau}$, if $\phi\tau$ is linear, and $\phi x^\tau \equiv 0^\omega$, if $\phi\tau = \omega$
(2.5) For the constants of T_J, the axioms governing these constants must be read by ϕ so that they become derivable in HA^ω. That leads to:
(2.5.1) $\phi(0) \equiv 0$ and $\phi(suc) \equiv suc$
(2.5.2) $\phi(\mathsf{R}_\tau) \equiv \mathsf{R}_{\phi\tau}$, if $\phi\tau$ is linear, and $\phi\mathsf{R}_\tau \equiv 0^\omega$, if $\phi\tau = \omega$
(2.5.3) $\phi(\mathsf{K}_{\sigma\tau}) \equiv \mathsf{K}_{\phi\sigma,\phi\tau}$, if $\phi\sigma, \phi\tau$ are linear,
$\qquad\qquad \equiv 0^\omega$ if $\phi\tau = \omega$,
$\qquad\qquad \equiv \mathsf{I}_{\phi\tau}$ if $\phi\sigma = \omega, \phi\tau$ is linear
(2.5.4) $\phi(\mathsf{S}_{\rho\sigma\tau}) \equiv \mathsf{S}_{\phi\rho,\phi\sigma,\phi\tau}$, if $\phi\rho, \phi\sigma, \phi\tau$ are linear
$\qquad\qquad \equiv 0^\omega$, if $\phi\tau = \omega$
$\qquad\qquad \equiv \mathsf{I}_{\phi\tau}$, if $\phi\tau$ is linear, $\phi\rho = \phi\sigma = \omega$
$\qquad\qquad \equiv \mathsf{I}_{\phi\sigma\to\phi\tau}$, if $\phi\sigma, \phi\tau$ are linear, $\phi\rho = \omega$
$\qquad\qquad \equiv \mathsf{I}_{\phi\rho\to\phi\tau}$, if $\phi\rho, \phi\tau$ are linear, $\phi\sigma = \omega$

(3) For the logical symbols of T, ϕ is a homomorphism, i.e.

(3.1) If $a = b$ is an equation of type τ in $L(T_J)$, then τ is linear, so $\phi\tau = \tau$, and we put $\phi(a = b) \equiv \phi a = \phi b$, again an equatiion of type τ.

(3.2) $\phi(\bot) \equiv \bot$, $\phi(A \wedge B) \equiv \phi(A) \wedge \phi(B)$, $\phi(A \to B) \equiv \phi(A) \to \phi(B)$

(3.3) ϕ *reads* the restricted quantifier $\forall x \in a$ in such a way that $\phi(J1)$ to $\phi(J4)$ are derivable in HA^ω and that the rule $\phi(J5)$ is admissible in HA^ω.

1.8.15   Theorem on readings of T_J in HA^ω

Readings ϕ of T_J in HA^ω are correct, i.e. they preserve derivability:

$$T_J \vdash A \quad \text{implies} \quad HA^\omega \vdash \phi(A)$$

The *proof* by induction on deductions is straight forward. The ϕ-readings of the axioms $(K), (S), (R)$ for the constants of T are derivable in HA^ω, also under context, due to (2.5) in the above definition. The logical axioms and rules of $T_\wedge$ stay correct under ϕ, because by (3), ϕ is a homomorphism on that part of the syntax. The ϕ-readings of the laws on the restricted quantifier are correct by part (3.3) of the above definition.

1.8.16   Corollary: $J\phi$-Interpretation theorem

For any reading ϕ of T_J, the composition $J\phi$ is a functional interpretation of HA^ω in itself:

$$HA^\omega \overset{J\phi}{\hookrightarrow} HA^\omega$$

Proof. Let $HA^\omega \vdash A$ with $A^J \equiv \exists v \forall w A_J[v, w]$. By the J-interpretation theorem 1.8.8, there is a tuple b of J-terms such that $T_J \vdash A_J[b, w]$. By the theorem on readings ϕ of T_J in HA^ω 1.8.15, $HA^\omega \vdash \phi(A_J[b, w])$, and that is

$$HA^\omega \vdash \phi(A_J)[\phi b, \phi w]$$

which proves the corollary.

We now turn to special readings and start with the simplest case.

1.8.17   The mr-reading ϕ_{mr}

In order to adapt the clause $(\to)^J$ to the mr-translation of implication, we simply have to cancel the restriction $\in Wvz$ and thus replace the restricted quantifier $\forall w \in Wvz$ by the unbounded universal quantifier $\forall w$. This is achieved by defining the mr-reading on types for all set types $\bar{\tau}$ by

$$\phi_{mr}\bar{\tau} = \omega$$

Therefore, for linear types τ, $\phi_{mr}(\forall x^\tau \in a) \equiv \forall x^\tau$: The restricted quantifier is mr-read as an unbounded universal quantifier. Clearly, under ϕ_{mr}, $J1$ to $J5$ hold in HA^ω.

1.8.18   Proposition: Equivalence of $J\phi_{mr}$ and mr

For $A \in L(HA^\omega)$ and $A^J \equiv \exists v' \forall w' \, A_J[v', w']$ with (possibly empty) tuples $\phi_{mr}(v', w') \equiv v, w$, we have

$$HA^\omega \vdash \forall w \phi_{mr}(A_J)[v, w] \leftrightarrow v \ mr \ A$$

and hence $HA^\omega \vdash A^{J\phi_{mr}} \leftrightarrow A^{mr}$, i.e. $J\phi_{mr} = mr$

Proof by subformula induction.

1. $\forall w \; \phi_{mr}(a = b) \equiv a = b \equiv v \ mr \ a = b$ with empty v, w.

Let $B^J \equiv \exists y' \forall z' B_J[y', z']$ with $\phi_{mr}(y', z') \equiv y, z$.

2. $\forall wz \; \phi_{mr}(A_J[v, w] \wedge B_J[y, z]) \leftrightarrow \forall w \phi_{mr} A_J[v, w] \wedge \forall z \phi_{mr} B_J[y, z]$
    $\leftrightarrow v \ mr \ A \wedge y \ mr \ B \leftrightarrow v, y \ mr \ (A \wedge B)$ by I.H.

3. $\forall vz \; \phi_{mr}(A \to B)_J[W, Y, v, z]$
    $\equiv \forall vz \; \phi_{mr}(\forall w \in W vz \, A_J[v, w] \to B_J[Yv, z])$
    $\leftrightarrow \forall v(\forall w \; \phi_{mr} \, A_J[v, w] \to \forall z B_J[Yv, z])$
    $\leftrightarrow \forall v(v \ mr \ A \to Yv \ mr \ B) \equiv Y \ mr \ (A \to B)$ by I.H.

4. $\forall w \; \phi_{mr} \forall u A_J[u, Vu, w] \equiv \forall uw \; \phi_{mr} A_J[u, Vu, w]$
    $\leftrightarrow \forall u \, Vu \ mr \ A[u] \equiv V \ mr \ \forall u A[u]$ by I.H.

5. $\forall w \; \phi_{mr} \exists u A_J[u, v, w] \equiv \forall w \; \phi_{mr} A_J[u, v, w]$
    $\leftrightarrow u, v \ mr \ A[u] \equiv v \ mr \ \exists u A[u]$

This proposition, together with the J-interpretation theorem 1.8.8 and corollary 1.8.16, yields an alternative proof of mr-realization of HA^ω in 1.6.7.

As the prototype of readings, we consider the $\wedge$-reading: The T_J-formula $\forall y \in a \, A[y]$ is to be read under $\wedge$ as $\forall x < lh(\phi_\wedge(a)) \, A[\phi_\wedge'(a)x]$, reading $\phi_\wedge(a)$ as a finite sequence. So, the $\wedge$-reading of a J-term a of set type $\bar{\tau}$ splits the term a into two components, its *length* $lh(\phi_\wedge(a)) : o$ and a *sequence* $\phi_\wedge'(a) : o \to \tau$ of which only the initial finite segment of length $lh(\phi_\wedge(a))$ is of interest. That leads to the following definition:

1.8.19   *The ∧-reading $\phi_\wedge$ of T_J*

$\phi_\wedge$ is defined on types by

$$\phi_\wedge(\bar{\tau}) = o, o \to \phi_\wedge(\tau),$$

on terms a of type $\bar{\tau}$ by

$$\phi_\wedge(a) \equiv lh(\phi_\wedge(a)) : o, \phi'_\wedge(a) : o \to \tau,$$

and on T_J-formulae $\forall y \in a\, A[y]$

$$\phi_\wedge(\forall y \in a\, A[y]) \equiv \forall x < lh(\phi_\wedge(a)]\, \phi_\wedge(A)[\phi'_\wedge(a)x]$$

The set forming constants are ∧-read as follows:

$\phi_\wedge(\{\ \})$: $lh(\phi_\wedge(\{c\})) = 1$; $\phi'_\wedge(\{c\}) \equiv \lambda z^o.\phi_\wedge(c)$

$\phi_\wedge(\cup)$: $lh(\phi_\wedge(a \cup b)) = lh(\phi_\wedge(a)) + lh(\phi_\wedge(b))$

$\phi'_\wedge(a \cup b)x \equiv \phi'_\wedge(a)x$ for $x < lh(\phi_\wedge(a))$

$\equiv \phi'_\wedge(b)(x \dot{-} lh(\phi_\wedge(a)))$ for $x \geq lh(\phi_\wedge(a))$

$\phi_\wedge(\bigcup)$:   For $a : \bar{\sigma}$, $f : \sigma \to \bar{\tau}$, hence $c := \bigcup\{fu^\sigma \mid u^\sigma \in a\} : \bar{\tau}$, put

$lh(\phi_\wedge(c)) = \Sigma_{y<lh(\phi_\wedge(a))}lh(\phi_\wedge(f)(\phi'_\wedge(a)y))$

$\phi'_\wedge(c)x \equiv \phi'_\wedge(f)(\phi'_\wedge(a)y)(x \dot{-} \Sigma_{z<y}lh(\phi_\wedge(f)(\phi'_\wedge(a)z))$

if $\Sigma_{z<y}lh(\phi_\wedge(f)(\phi'_\wedge(a)z)) \leq x < \Sigma_{z\leq y}lh(\phi_\wedge(f)(\phi'_\wedge(a)z))$

By lemmata 1.1.20 and 1.1.21, it is easily seen that these ∧-readings of the set-forming constants satisfy the ∧-readings of $J1$ to $J5$. Moreover, the ∧-reading leads to the following characterization of the ∧-translation:

1.8.20   *Proposition: Equivalence of $J\phi_\wedge$ and $\wedge$*

For $A^J \equiv \exists v' \forall w'\, A_J[v', w']$ with tuples $\phi_\wedge(v', w') \equiv v, w$, we have

$$HA^\omega \vdash \phi_\wedge(A_J)[v, w] \leftrightarrow A_\wedge[v, w]$$

and hence $HA^\omega \vdash A^{J\phi_\wedge} \leftrightarrow A^\wedge$, i.e. $J\phi_\wedge = \wedge$.

Proof by subformula induction.

1. $\phi_\wedge(a = b) \equiv a = b \equiv (a = b)_\wedge$
2. $\phi_\wedge((A \wedge B)_J)[v, y, w, z] \equiv \phi_\wedge(A_J)[v, w] \wedge \phi_\wedge(B_J)[y, z]$
   $\leftrightarrow A_\wedge[v, w] \wedge B_\wedge[y, z] \equiv (A \wedge B)_\wedge[v, y, w, z]$ by I.H.
3. With $\phi_\wedge W' \equiv lh(\phi)_\wedge W', \phi'_\wedge W' \equiv X, W$, we obtain
   $\phi_\wedge((A \to B)_J)[W', Y', v', z']$
   $\equiv \phi_\wedge(\forall w \in W'v'z'\, A_J[v', w] \to B_J[Y'v', z'])$
   $\leftrightarrow (\forall x < Xvz\, A_\wedge[v, Wvzx] \to B_\wedge[Yv, z])$ by I.H.
   $\equiv (A \to B)_\wedge[X, W, Y, v, z]$

4. $\phi_\wedge((\forall u\, A[u])_J)[v, w] \equiv (\phi_\wedge(A[u])_J)[vu, u, w]$
   $\qquad \leftrightarrow A_\wedge[vu, u, w]$ by I.H.
5. $\phi_\wedge((\exists u\, A[u])_J)[v, w] \equiv \phi_\wedge((A[u]_J)[u, v, w]) \leftrightarrow A_\wedge[u, v, w]$ by I.H.
   $\qquad \equiv (\exists u\, A[u])_\wedge[v, w]$

Again, this proposition, together with theorem 1.8.8 and corollary 1.8.16, yields an alternative proof of the $\wedge$- interpretation theorem 1.2.7.

There are further readings of J "between" ϕ_{mr} and $\phi_\wedge$. For them, we refer to the *degree* of a type.

1.8.21   Recursive definition of the degree $d(\tau)$ of J-types τ and of pure linear types

1. $d(o) = 0$
2. $d(\sigma \to \tau) = max(d(\sigma) + 1, d(\tau))$
3. $d(\bar\tau) = d(\tau)$

Pure linear types are usually denoted by natural numbers:

1. $0 :\equiv o$
2. $n + 1 :\equiv n \to o$

Clearly, the pure linear type n has degree n.

1.8.22   Coding of pairs in absence of product types

For the rest of this section, we assume the existence of *pair coding functionals* for linear types σ, τ with $d(\sigma) \leq d(\tau)$:

$$\langle\, ,\, \rangle_{\sigma\tau} : \sigma \to \tau \to \tau$$

together with their inverses $j_0 : \sigma \to \tau$, $j_1 : \tau \to \tau$ such that

$$j_i\langle a_0, a_1\rangle_{\sigma\tau} = a_i \quad \text{for } i = 0, 1$$

Such codings of pairs are certainly definable in extensional versions of HA^ω. For $\sigma = \tau = o\, \langle\, ,\, \rangle_{oo}$ is a well known primitive recursive function.

For any degree n, we now define a reading ϕ_n corresponding to ϕ_{mr} for "small" types of degree $\leq n$ and closer to $\phi_\wedge$ for "large" types of degree $> n$:

1.8.23   Definition of the reading $\phi_n : T_J \to HA^\omega$ for $n \in \mathbb{N}$

If $d(\tau) \leq n$, we put $\phi_n(\bar\tau) = \omega$ and hence for linear types τ

$$\phi_n(\forall x^\tau \in a\, B) \equiv \forall x^\tau\, \phi_n(B)$$

Then $J1$ to $J5$ hold under ϕ_n as they hold under ϕ_{mr} (cf. 1.8.17).
Now let $d(\tau) > n$. Then we put $\phi_n(\bar{\tau}) = n \to \phi_n(\tau)$ and for linear types τ:

If $c : \tau$ then $\phi_n(\{c\}) = \lambda z^n . \phi_n(c)$

If $a, b : \bar{\tau}$ then let $z^n = \langle x^o, y^n \rangle$ and put

$\quad \phi_n(a \cup b)z^n \qquad = \phi_n(a)y^n$ if $x^o = 0$

$\qquad\qquad\qquad\qquad = \phi_n(b)y^n$ if $x^o \neq 0$

If $f : \sigma \to \bar{\tau}$ and $a : \bar{\sigma}$ then

$\quad$ for $d(\sigma) \leq n$,   let $z^n = \langle x^\sigma, y^n \rangle$ and put

$\quad \phi_n(\bigcup\{fu^\sigma \mid u^\sigma \in a\})z = \phi_n(f)x^\sigma y^n$

$\quad$ for $d(\sigma) > n$,   let $z^n = \langle y^n, x^\sigma \rangle$ and put

$\quad \phi_n(\bigcup\{fu^\sigma \mid u^\sigma \in a\})z = \phi_n(f)(\phi_n(a)x^\sigma)y^n$

and finally

$$\phi_n(\forall u^\tau \in a \ B[u^\tau]) \equiv \forall x^n \phi_n(B)[\phi_n(a)x^n]$$

1.8.24   *Proposition: ϕ_n models T_J in HA^ω*

For any n, ϕ_n is a (correct) reading of T_J in HA^ω, i.e. $J1$ to J_5 hold in HA^ω under ϕ_n.

Proof. For $d(\tau) \leq n$, $J1$ to $J5$ hold under ϕ_n as they hold under ϕ_{mr}. So let $d(\tau) > n$. Then in HA^ω are derivable:

$\phi_n(J1) \quad \forall x^n \phi_n(B)[\lambda x^n . \phi_n(a^\tau)x^n] \to \phi_n(B)[\phi_n(a^\tau)]$

$\phi_n(J2) \quad \forall z^n \phi_n(C)[\phi_n(a \cup b)z^n] \to$
$\qquad\qquad \forall x^o y^n \phi_n(C)[\phi_n(a \cup b)\langle x^o, y^n \rangle] \to$
$\qquad\qquad \forall y^n \phi_n(C)[\phi_n(a)y^n] \wedge \forall y^n \phi_n(C)[\phi_n(b)y^n]$

$\phi_n(J3) \quad for \ d(\sigma) \leq n :$
$\qquad\qquad \forall z^n \phi_n(C)[\phi_n(f)(j_1 z)(j_2 z)] \to$
$\qquad\qquad \forall x^\sigma \forall y^n \phi_n(C)[\phi_n(f)x^\sigma y^n]$

$\phi_n(J3) \quad for \ d(\sigma) > n :$
$\qquad\qquad \forall z^n \phi_n(C)[\phi_n(f)(\phi_n(a)(j_1 z^n))(j_2 z^n)] \to$
$\qquad\qquad \forall x^n \forall y^n \phi_n(C)[\phi_n(f)(\phi_n(a)x^n)y^n]$

The ϕ_n-readings of $J4$ and $J5$ are straight forward and are left to the reader.

By corollary 1.8.16 and this proposition, the functional translation $J\phi_n$ is, for any natural number n, a functional interpretation of HA^ω in itself:

$$HA^\omega \overset{J\phi_n}{\hookrightarrow} HA^\omega$$

In analogy to $J\phi_{mr} = mr$ in 1.8.18 and $J\phi_\wedge = \wedge$ in 1.8.20, we give a direct formulation of the functional translation $n = J\phi_n$:

1.8.25 *Recursive definition of n-translations*

Given a natural number n, to any formula A of $L(HA^\omega)$, the *n-translation* assigns a formula

$$A^n \equiv \exists v \forall w\, A_n[v, w] \equiv \exists v \forall w^{\leq n} w^{>n} A_n[v, w^{\leq n}, w^{>n}],$$

where $w^{\leq n}$ and $w^{>n}$ stand for the subtuples of variables of w of type degree $\leq n$ and $> n$, respectively, which is again a formula of HA^ω. A^n is defined recursively on $L(HA^\omega)$ like the J-translation in definition 1.8.6 (with J replaced by n), with the exception of the clause for implication which now runs:

With A^n as above and $B^n \equiv \exists y \forall z\, B_n[y, z]$,

$$(\rightarrow)^n \quad (A \rightarrow B)^n \equiv \exists W Y \forall v z (\forall w^{\leq n} x^n\, A_n[v, w^{\leq n}, W v z x^n] \rightarrow B_n[Y v, z])$$

Clearly, the n-translation is, like $J\phi_n$ a functional translation from $L(HA^\omega)$ into $L(HA^\omega)$.

1.8.26 *Proposition: n- and $J\phi_n$-translation coincide*

For any n, the recursive definition of the n-translation coincides with the $J\phi_n$-translation:

$$A^n \equiv A^{J\phi_n}$$

Proof by subformula induction. Since ϕ_n is the identity map on $L(HA^\omega)$, and n and J coincide on connectives other than implication, it suffices to show the induction step for implication (we again split the tuple w into the subtuples $w^{\leq n}$ and $w^{>n}$ according to their type degree):

$\phi_n(A \rightarrow B)^J \equiv$
$\phi_n(\exists W' W Y \forall v z (\forall w^{\leq n}, w^{>n} \in W' v z, W v z\, A_J[v, w^{\leq n}, w^{>n}] \rightarrow B_J[Y v, z]))$
$\equiv \exists W Y \forall v z (\forall w^{\leq n} x^n \phi_n(A_J)[v, w^{\leq n}, W v z x^n] \rightarrow \phi_n(B_J)[Y v, z])$
$\equiv \exists W Y \forall v z (\forall w^{\leq n} x^n A_n[v, w^{\leq n}, W v z x^n] \rightarrow B_n[Y v, z])$ by I.H.
$\equiv (A \rightarrow B)^n$

Together with corollary 1.8.16, this immediately yields:

1.8.27 *Corollary: n-Interpretation theorem*

For any n, HA^ω is n-interpretable in HA^ω:

$$HA^\omega \overset{n}{\hookrightarrow} HA^\omega$$

1.9 Interpretations of classical arithmetic

In the context of functional translations and interpretations, it seems adequate to formulate classical theories in the negative fragment, i.e. in the $\{\exists, \vee\}$free fragment of predicate logic. In this fragment, stability is the one logical principle that extends intuitionistic to classical logic. It is an elementary theorem of intuitionistic logic that in intuitionistic theories Th stability of arbitrary $\{\exists, \vee\}$free formulae is derivable from the stability of their atomic formulae:

1.9.1 *Stability lemma*

$$Th + \{\neg\neg P \to P \mid P \text{ atomic }\} \vdash \neg\neg A \to A$$

for all $\{\exists, \vee\}$free formulae A in $L(Th)$.

The proof of this lemma (by subformula induction) is left to the reader.

1.9.2 *Description of classical arithmetical systems*

HA as well as T_0 prove $\neg\neg s = t \to s = t$ for all equations $s = t$ within their respective language; moreover, we consider disjunction $\vee$ as a defined symbol by 1.2.2 already in HA. Therefore, classical Peano arithmetic PA may be taken to be the $\exists$free fragment of HA, and by lemma 1.3.6, T_0 and $T_{\wedge 0}$ as well as I-T and I-$T_\wedge$ are themselves already classical functional theories. If we denote by $Stab(=)$ the axiom schema for stability of equations of arbitrary type

$$Stab(=) :\equiv \{\neg\neg a = b \to a = b \mid a, b : \tau, \ \tau \text{ a type}\},$$

the classical versions of the theories T and $T_\wedge$ are defined as

$$T^c :\equiv T + Stab(=) \quad \text{and} \quad T^c_\wedge :\equiv T_\wedge + Stab(=)$$

Finally, Peano arithmetic in all finite types PA^ω is defined as the $\exists$free fragment of HA^ω, extended by the schema $Stab(=)$. Its type o fragment PA^ω_0 is simply the $\exists$free fragment of HA^ω_0, and its intensional extension I-PA^ω is the $\exists$free fragment of I-HA^ω.

By lemma 1.9.1, these theories satisfy stability for their full language.

The D- and $\wedge$-interpretation theorems in sections 1.3 and 1.2 immediately yield the corollaries:

1.9.3   *D-* and ∧-*interpretation theorems for classical arithmetic*

$$PA \overset{D}{\hookrightarrow} T_0$$

$$PA_0^\omega \overset{D}{\hookrightarrow} T_0 \quad and \quad I\text{-}PA^\omega \overset{D}{\hookrightarrow} I\text{-}T$$

$$PA^\omega \overset{\wedge}{\hookrightarrow} T_\wedge^c, \quad in\ fact\ even \quad HA^{\omega+} + Stab(=) \overset{\wedge}{\hookrightarrow} T_\wedge^c$$

The D-interpretation theorems are immediate from 1.3.9, because PA, PA_0^ω, and $I\text{-}PA^\omega$ are subtheories of the corresponding intuitionistic theories; for the ∧-interpretation theorems, the proofs in 1.2.7 and 1.2.13 must be extended only to the schema $Stab(=)$ which is translated by itself.

In studying the strength of translations D and $\wedge$ in PA^ω, we are in a position different from the constructive case, because the translation $A^I \equiv \exists v \forall w A_I[v, w]$ is in general, even for $\exists$free A, not a formula of PA^ω. Instead of A^I, we consider its *negative version* $A^{I-} \equiv \neg\forall v \neg\forall w A_I[v, w]$.

1.9.4   *Negative version* ⁻ *on* $L(HA^\omega)$, *definition and lemma*

The **negative version** A^- of a formula $A \in L(HA^\omega)$ is obtained from A by replacing any non-empty tuple of existential quantifiers $\exists u$ by $\neg\forall u\neg$. ⁻ is thus the identity map on $L(PA^\omega)$, and ⁻ is idempotent on all of $L(HA^\omega)$ Whereas by definition PA^ω is a subtheory of $HA^\omega + Stab(=)$,

$$PA^\omega \prec HA^\omega + Stab(=),$$

in the reverse direction, ⁻ is an embedding of $HA^\omega + Stab(=)$ in PA^ω which is easily seen by induction on deductions:

Lemma

$$HA^\omega + Stab(=) \vdash A \quad implies \quad PA^\omega \vdash A^-$$

In HA^ω, for quantifier free A, the I-translation of $\forall x \exists y A[x, y]$ is $\exists Y \forall x A[x, Yx]$, for $I = D$ as well as for $I = \wedge$. Thus,

$$B \to B^I \text{ with } B \equiv \forall x \exists y\ A[x, y]$$

is an axiom of choice with qf matrix A for both I in question. Combining these translations with the negative version complicates the situation.

1.9.5 Axioms of choice in classical context; theories $PA^{\omega*}$ and $PA^{\omega+}$

For $A \in L(T)$ and non-empty tuples x, y of variables of arbitrary type, let the **quantifier free axiom of choice** be the schema

$(qf - AC)$ $\quad \forall x \neg \forall y \neg A[x, y] \to \neg \forall Y \neg \forall x \, A[x, Yx],$

which is $(AC)^-$ with quantifier free matrix A.

Up to a double negation which is irrelevant in PA^{ω} due to the stability lemma 1.9.1, $(qf - AC)$ is of the form

$$B \to B^{D-} \text{ with } B \equiv (\forall x \exists y \, A[x, y])^-$$

Similarly, for $A \in L(T_\wedge)$ and tuples x, y as above, let

$(qf - ARC)$ $\quad \forall x \neg \forall y \neg A[x, y] \to \neg \forall S, Y \neg \forall x \neg \forall s < Sx \neg A[x, Ysx]$

This **quantifier free axiom of restricting choice** is literally of the form

$$B \to B^{\wedge-} \text{ with } B \equiv (\forall x \exists y \, A[x, y])^-$$

We define

$$PA^{\omega*} :\equiv PA^{\omega} + (qf - AC) \quad and \quad PA^{\omega+} :\equiv PA^{\omega} + (qf - ARC)$$

1.9.6 Lemma

For $A, B, A[u]$ $\exists$free, u a - possibly empty - tuple of variables, PA^{ω} proves:
1. $A^{\wedge-} \leftrightarrow A$ for $A \in L(T_\wedge)$
2. $(A \wedge B)^{\wedge-} \leftrightarrow A^{\wedge-} \wedge B^{\wedge-}$
3. $(A \to B)^{\wedge-} \to A^{\wedge-} \to B^{\wedge-}$
4. $(\forall u \neg A[u])^{\wedge-} \to \forall u \neg (A[u]^{\wedge-})$; furthermore
5. $PA^{\omega+} \vdash \forall u \neg (A[u]^{\wedge-}) \to (\forall u \neg A[u])^{\wedge-}$

Proof. 1. to 3. are straightforward and follow from the corresponding statements in 1.2.12.

4.: For $A[u]^{\wedge} \equiv \exists v \forall w \, A_\wedge[u, v, w]$, we have

$$(\forall u \neg A[u])^{\wedge-} \equiv (\exists XW \forall uv \neg \forall x < Xuv \, A_\wedge[u, v, Wxuv])^-$$

This implies $(\forall uv \exists XW \exists x < X \neg A_\wedge[u, v, Wx])^-$ which is equivalent to

$$(\forall uv \exists w \neg A_\wedge[u, v, w])^- \equiv \forall u \neg (A[u]^{\wedge-})$$

5.: By $(qf - ARC)$, this last formula implies

$$(\exists XW \forall uv \exists x < Xuv \neg A_\wedge[u, v, Wxuv])^-$$

which up to a double negation is $(\forall u \neg A[u])^{\wedge-}$.

1.9.7   Characterization theorem for the $\wedge$-translation on PA^ω

1. The theories $PA^{\omega+}$ and $PA^\omega + \{A \leftrightarrow A^{\wedge-}\}$ are equivalent:

$$PA^\omega + (qf - ARC) \quad \equiv \quad PA^\omega + \{A \leftrightarrow A^{\wedge-}\}$$

2. $PA^{\omega+}$ is $\wedge$-maximal relative $T_\wedge^c$: Any formula $\wedge$-interpretable in $T_\wedge^c$ is derivable in $PA^{\omega+}$.

Proof of 1. The direction from right to left follows from the observation in 1.9.5 that axioms $(qf - ARC)$ are instances of the schema $\{A \to A^{\wedge-}\}$. For the other direction, we follow an idea of Kreisel [29] - which he applies to the Dialectica translation - and exploit the fact that every formula A of $L(PA^\omega)$ possesses a prenex normal form $B \equiv \forall v_1 \neg \forall v_2 \neg ... \neg \forall v_n \neg C[v]$ with - possibly empty - tuples of variables $v_1, ..., v_n$ and $C[v] \in L(T_\wedge)$. For this B, we have

(1) $PA^\omega \vdash A \leftrightarrow B$. Therefore, by the $\wedge$-interpretation theorem 1.9.3,
(2) $PA^\omega \vdash (A \leftrightarrow B)^{\wedge-}$, which by 2. and 3. in 1.9.6 implies
(3) $PA^\omega \vdash A^{\wedge-} \leftrightarrow B^{\wedge-}$

By n applications of 4. and 5. in 1.9.6, we obtain

(4) $PA^{\omega+} \vdash B^{\wedge-} \leftrightarrow B$

Equivalences (1), (4), and (3) together imply

$$PA^{\omega+} \vdash A \leftrightarrow A^{\wedge-}$$

So, for the classical theory PA^ω, the classical prenex normal form is the adequate normal form to allow for a simple proof of a characterization of the $\wedge$-translation.

Proof of 2. Let A be $\wedge$-interpretable in $T_\wedge^c$. Then $A^\wedge$ is derivable in $HA^\omega + Stab(=)$, and by lemma 1.9.4, $A^{\wedge-}$ is derivable in PA^ω. Thus, by 1., A is derivable in $PA^{\omega+}$.

1.9.8   Proposition

$PA^{\omega+}$ is a subtheory of $HA^{\omega+} + Stab(=)$:

$$PA^{\omega+} \prec HA^{\omega+} + Stab(=)$$

Proof. Since $PA^\omega \prec HA^\omega + Stab(=)$, it remains to show for $\exists$free A:
(1) $HA^{\omega+} + Stab(=) \vdash A \leftrightarrow A^{\wedge-}$
For $A^\wedge \equiv \exists v \forall w A_\wedge[v, w]$, we have by intuitionistic logic
$$\vdash \neg\neg A^\wedge \leftrightarrow \neg \forall v \neg \forall w A_\wedge[v, w,] \equiv A^{\wedge-}.$$
Since A is $\exists$free, the stability lemma 1.9.1 implies

$HA^\omega + Stab(=) \vdash A \leftrightarrow \neg\neg A$.

However, $HA^{\omega+} \vdash A \leftrightarrow A^\wedge$ and hence also $\vdash \neg\neg A \leftrightarrow \neg\neg A^\wedge$

Putting these three equivalences together yields (1) and thus the proposition.

Theorem 1.9.3 and proposition 1.9.8 immediately imply as a corollary:

1.9.9 Extended ∧-interpretation theorem

$$PA^{\omega+} \stackrel{\wedge}{\hookrightarrow} T_\wedge^c$$

As in section 1.3, corresponding results also hold for the Dialectica interpretation of PA_0^ω as well as of $I\text{-}PA^\omega$.

1.9.10 Proposition

For $Th \equiv PA_0^\omega$ as well as for $Th \equiv I\text{-}PA^\omega$, if $A \in L(Th)$, then

$$Th \vdash A^{D-} \leftrightarrow A^{\wedge-}$$

On the basis of PA_0^ω as well as of $I\text{-}PA^\omega$, the schemata $(qf - AC)$ and $(qf - ARC)$ are equivalent.

Proof. The equivalence in the first statement is the negative version of the equivalence in proposition 1.3.11. Therefore, the statement follows by lemma 1.9.4, adapted to the theories Th.

The second statement follows from the first, because, as remarked in 1.9.5, axioms $(qf-AC)$ and $(qf-ARC)$ are of the form $B \to B^{D-}$ and $B \to B^{\wedge-}$ respectively, for the same B.

As corollaries to the characterization theorem 1.9.7 and the extended interpretation theorem 1.9.9, this proposition implies:

1.9.11 Characterization and extended interpretation theorems for the Dialectica translation

(1) For $Th = PA_0^\omega$ as well as for $Th = I\text{-}PA^\omega$, the theories Th^* and $Th + \{A \leftrightarrow A^{D-}\}$ are equivalent:

$$PA_0^{\omega*} \equiv PA_0^\omega + \{A \leftrightarrow A^{D-}\} \quad and \quad I\text{-}PA^{\omega*} \equiv I\text{-}PA^\omega + \{A \leftrightarrow A^{D-}\}$$

(2) Any formula D-interpretable in T_0 is derivable in $PA_0^{\omega*}$.

(3) Any formula D-interpretable in $I\text{-}T$ is derivable in $I\text{-}PA^{\omega*}$.

(4) $\qquad\qquad PA_0^{\omega*} \stackrel{D}{\hookrightarrow} T_0 \quad and \quad I\text{-}PA^{\omega*} \stackrel{D}{\hookrightarrow} I\text{-}T$

Proof. (1), with $+$ instead of $*$ and $\wedge$ instead of D, follows from 1. in 1.9.7

by restricting the language of PA^ω to its type o fragment respectively by extending it by higher type equality functionals. By proposition 1.9.10, these variants are equivalent to (1). By the same argument, (2) and (3) follow from 2. in 1.9.7.

Claims (4) extend theorem 1.9.3 only by the D-interpretation of $(qf - AC)$ which is of the form $B \to B^{D-}$, and that is D-interpreted as follows:

Let $B^D \equiv \exists v \forall w B_D[v, w]$ and $B^{D-} \equiv \neg \forall y \neg \forall z B_D[y, z]$.

Then $B^{D-D} \equiv \exists Y \forall Z \neg \neg B_D[YZ, Z(YZ)]$ and finally

$$(B \to B^{D-})^D \equiv \exists W Y \forall v Z (B_D[v, WvZ] \to \neg \neg B_D[Yvz, Z(Yvz)])$$

D-interpreting functionals W, Y are given by $YvZ = v$ and $WvZ = Zv$. That proves (4).

These results complete the Dialectica interpretation of $PA_0^{\omega*}$ and of $I\text{-}PA^{\omega*}$. They correspond to results in section 1.3, with $(qf - AC)$ taking the place of the schemata (M), (IP), and (AC). Concerning the translations D and $\wedge$ on PA^ω proper, some details remain to be settled, some rest unsolved.

1.9.12   *Proposition*

Let $(qf - AC)_0$ denote the schema $(qf - AC)$, restricted to type o formulae.

(1) $PA^{\omega*} \vdash (qf - ARC)$

(2) $PA^{\omega+} \vdash (qf - AC)_0$

(3) $PA^\omega \nvdash (qf - AC)_0$

(4) $PA^\omega \nvdash (qf - ARC)$

(5) $PA^{\omega+} \vdash (qf - AC)$ if and only if $(qf - AC)$ is $\wedge$-interpretable in $T_\wedge^c$.

Proof. (1) $(\exists Y' \forall x A[x, Y'x])^-$ implies $(\exists S, Y \forall x \exists s < SxA[x, Ysx])^-$, simply by putting $Sx = 1$ and, given Y', $Ysx = Y'x$. Therefore, any axiom $(qf - AC)$ implies the corresponding axiom $(qf - ARC)$.

(2) is an immediate consequence of proposition 1.9.10.

(3) Let T denote Kleene's primitive recursive T-predicate. For any numeral e, the negative version of

$$\forall x \exists y \, \mathsf{T}exy \to \exists Y \forall x \, \mathsf{T}ex(Yx)$$

is an instance of $(qf - AC)_0$. Now let e be an index of a total recursive function which is not provably recursive in PA^ω. In the model NF of primitive recursive functionals in normal form (cf. definition 1.5.22), we have

$$NF \models \forall x \exists y \, \mathsf{T}exy \quad \text{but for no } Y \quad NF \models \forall x \mathsf{T}ex(Yx)$$

Therefore $NF \not\models (qf - AC)_0$, and (3) follows, since NF is a model of PA^ω. (4) is immediate from (3) and proposition 1.9.10.
(5) is an application of the characterization theorem 1.9.7,2 and the extended $\wedge$-interpretation theorem 1.9.9.

By (1) of this proposition, the sequence of theories

$$PA^{\omega*} \qquad PA^{\omega+} \qquad PA^\omega$$

is of decreasing strength, and by (4), the second theory is properly stronger than the last. We conjecture that also $PA^{\omega+} \not\vdash (qf - AC)$.

Lemma 1.9.6, with $\wedge$ replaced by D and $PA^{\omega+}$ replaced by $PA^{\omega*}$, still holds for the full language of PA^ω. This implies

$$PA^{\omega*} \vdash A \leftrightarrow A^D$$

for all prenex A, even for all A generated from formulae of $L(T)$ by conjunction $\wedge$ and the combination $\forall u\neg$ with u any tuple of variables. Due to the lack of a D-interpretation theorem for PA^ω, it remains an open problem whether $PA^{\omega*} \vdash A \leftrightarrow A^D$ for all $A \in L(PA^\omega)$. Again, our conjecture is negative.

1.10 Extensionality and majorizability

In section 1.2, we $\wedge$-interpreted the neutral or combinatorial version HA^ω of Heyting arithmetic in finite types and did not consider its extensional versions $WE - HA^\omega$ and $E - HA^\omega$. This will be done now.
In this section, it will suffice to consider *linear types* only; every linear type τ is of a form $\tau = \sigma \to o$ with σ a tuple of linear types. In this section, the term *type* will henceforth refer to linear types.

1.10.1 *Weakly extensional and extensional extensions of* HA^ω

We take over the axiom of type-extensionality $(t - ext)$ from 0.1.14 and specify the rule $(T - EXT)$ to side formulae from $L(T_\wedge)$:

$(T - EXT)$ $\quad A \to fx = gx \vdash A \to f = g$
$\qquad\qquad$ for x not free in A, f, g, and A a formula of $L(T_\wedge)$.

Then let

$$WE - HA^{\omega(+)} :\equiv HA^{\omega(+)} + (T - EXT)$$
$$E - HA^\omega :\equiv HA^\omega + (t - ext)$$

Clearly, the $\wedge$-interpretation theorems 1.2.7 and 1.2.13 extend to weak extensionality, because the rule $(T - EXT)$ is $\wedge$-translated by itself:

1.10.2   Corollary to theorem 1.2.13: $\wedge$-interpretation of $WE - HA^{\omega+}$

The theory $WE - HA^{\omega+}$ is $\wedge$-interpretable in $T_\wedge + (T - EXT)$:

$$WE - HA^{\omega+} \overset{\wedge}{\hookrightarrow} T_\wedge + (T - EXT)$$

Following Howard [23], it will now be shown that this result does not extend to $E - HA^\omega$. In 0.1.14.4, the following special case of $(t - ext)$ is mentioned:

(E_2) $\forall Y^2 u^1 v^1 (\forall x^o (ux = vx) \to Yu = Yv)$

This formula is $\wedge$-translated (up to logical equivalence):

1.10.3 $\wedge$-translation of (E_2)

$(E_2)^\wedge$ $\exists X \forall Y uv (\forall x < XY uv(ux = vx) \to Yu = Yv)$

Howard [23] introduces the concept of *majorization / majorizability* to separate the primitive recursive functionals from functionals D-interpreting (E_2) and proves:
(*1) (E_2) is not D-interpretable by *majorizable* functionals;
(*2) Any primitive recursive functional is *majorizable*.
As (E_2) is a type o formula, (*1) implies that (E_2) is also not $\wedge$-interpretable.

1.10.4   Recursive definition of the relation f^* majorizes f, f^* maj f for functionals $f^*, f : \tau$ by recursion on linear types τ.

1. For $f^*, f : o$, f^* *maj* f iff $f^* \geq f$.
2. For $f^*, f : \sigma \to \tau$, f^* *maj* f iff, for all g^*, g of type σ, g^* *maj* g implies $f^* g^*$ *maj* fg.

A functional $f : \tau$ is **majorizable**, if there is a functional $f^* : \tau$ that majorizes f.

1.10.5   Extending the tuple notation

We extend the tuple notation of 0.1.13 to the predicate *maj*: If σ is a type tuple $\sigma_1, ..., \sigma_n$, and g^* and g are tuples of functionals $g_1^*, ..., g_n^*$ and $g_1, ..., g_n$ of type tuple σ , then g^* *maj* g stands for the conjunction g_1^* *maj* $g_1 \wedge ... \wedge g_n^*$ *maj* g_n.

1.10.6   *Lemma*

Let τ, ρ be types and let σ be a type tuple of length n such that $\tau = \sigma \to \rho$, and let $f^*, f : \tau$.
(1)   *f^* maj f iff for all $g^*, g : \sigma$ g^* maj g implies $f^* g^*$ maj fg.*
In particular:
(2)   *f^* maj f and g^* maj g implies $f^* g^*$ maj fg,*
and in case $\rho = o$,
(3)   *f^* maj f iff for all $g^*, g : \sigma$ g^* maj g implies $f^* g^* \geq fg$.*

Proof of (1) by induction on n. For $n = 0$, g^*, g are empty, and premise and conclusion coincide. For the induction step, let g^*, g be n-tuples and let h^*, h be single functionals. Then by I.H.,
f^* *maj* f iff $\forall g^* g (g^*$ *maj* $g \to \forall h^* h(h^*$ *maj* $h \to f^* g^* h^*$ *maj* $fgh))$
which is equivalent to $\forall g^* h^* gh(g^*, h^*$ *maj* $g, h \to f^* g^* h^*$ *maj* $fgh)$.
(2) is the direction from left to right of (1), and (3) is the case $\rho = o$ of (1).

1.10.7   *Lemma*

Let f', f be functionals of type $o \to \sigma \to o$, σ a type tuple. If $f't$ maj ft for all functionals $t : o$, then $f^* := \lambda x^o y^\sigma . Max\{f'iy \mid i \leq x\}$ majorizes f.

Proof. For all $t^* \geq t$, g^* *maj* g (of type tuple σ),

$$f^* t^* g^* = Max\{f'ig^* \mid i \leq t^*\} \geq f'tg^* \geq ftg$$

by (3) of the previous lemma, and therefore f^* *maj* f for the same reason.

1.10.8   *Theorem: Majorizability of primitive recursive functionals*

Any primitive recursive functional, even any functional of I-T, is majorized by a primitive recursive functional.

Proof by induction on the construction of functionals.
1. By (2) of lemma 1.10.6, the class of majorizable functionals is closed under application.
2. 0, suc, K , and S are self majorizing, because
$0 \geq 0$, $t^* \geq t$ implies $suct^* \geq suct$,
a^* *maj* a implies $Ka^* b^* = a^*$ *maj* $a =$ Kab, and
a^*, b^*, c^* *maj* a, b, c implies $b^* c^*$ *maj* bc by (2) of 1.10.6 and hence
$Sa^* b^* c^* = a^* c^* (b^* c^*)$ *maj* $ac(bc) =$ Sabc, again by (2) of 1.10.6.
3. Since, for $a, b : \tau$, E$_\tau ab \leq 1$, by (3) of 1.10.6, $\lambda u^\tau v^\tau .1$ *maj* E$_\tau$.

4. Let a^*, b^* maj a, b and put $f' := Ra^*b^*, f := Rab$. Then
$f'0 = b^*$ maj $b = f0$, and
$f'n$ maj fn implies $f'(suc\,n) = a^*n(f'n)$ maj $an(fn) = f(suc\,n)$
by (2) of 1.10.6. So, by induction on n, $f't$ maj ft for all $t : o$, and by
lemma 1.10.7,

$$R^* := \lambda abx^o y.Max\{Rabiy \mid i \leq x\} \ maj \ R,$$

and R^* is primitive recursive. That completes the proof.

It remains to show (*1) in 1.10.3 which will be done in the form: *Majorizable
functionals do not satisfy* $(E_2)^\wedge$.

1.10.9   Definition: Bounded classes $M(\sigma; k)$

For $\sigma = \rho \to o$ with ρ a type tuple, and k a numeral, put

$$M(\sigma; k) := \{b : \sigma \mid \forall z^\rho \ bz \leq k\} = \{b : \sigma \mid \lambda z^\rho.k \ maj \ b\}$$

If σ is a type tuple $\sigma_1, ..., \sigma_r$, then let

$$M(\sigma; k) := M(\sigma_1; k) \times ... \times M(\sigma_r; k)$$

1.10.10   Lemma

*Let $\tau = \sigma \to o$, σ a type tuple, and let k be a numeral.
If $a : \tau$ is majorizable, then there is a numeral m such that $ab \leq m$ for all
$b \in M(\sigma; k)$.*

Proof. Let $\sigma = \sigma_1, ..., \sigma_r$ and each $\sigma_i = \rho_i \to o$ (ρ_i a type-tuple). Then,
for $i = 1, ..., r$, any $b_i \in M(\sigma_i; k)$ is majorized by $\lambda z_i^{\rho_i}.k$. If $a : \sigma \to o$ is
majorized by, say, a^*, then, by lemma 1.10.6 (3),

$$m := a^*(\lambda z_1.k)...(\lambda z_r.k) \geq ab_1...b_r \equiv ab$$

1.10.11   Theorem: Non-majorizability of functionals $\wedge$-interpreting extensionality

Functionals X satifying $(E_2)^\wedge$ are not majorizable.

Proof. Consider the following primitive recursive families of functions
$\lambda n.u_n : o \to 1$ and functionals $\lambda n.Y_n : o \to 2$ defined by
$u_n x = 0$ for $x < n$ and $u_n x = 1$ for $x \geq n$
$Y_n u = 1$ if $\forall x < n \ ux = 0$ and $un = 1$; $Y_n u = 0$ otherwise.
Clearly, for all n, $u_n \in M(1; 1)$ and $Y_n \in M(2; 1)$.

Assume X to satisfy $(E_2)^\wedge$ (cf. 1.10.3). Then X will in particular satisfy the matrix of $(E_2)^\wedge$ for

$$Y = Y_n, u = u_n, v = 0^1 \text{ for all } n : o, \text{ i.e.}$$

$$\forall n (\forall x < XY_n u_n 0^1 \quad u_n x = 0 \rightarrow Y_n u_n = 0)$$

But $Y_n u_n = 1$. Therefore $\forall n \exists x < XY_n u_n 0^1 \quad u_n x = 1$.
However, $u_n x = 1$ holds only for $n \leq x$. Therefore finally

$$\forall n \quad n < XY_n u_n 0^1,$$

and, since all Y_n are in $M(2;1)$, all u_n in $M(1;1)$ (and 0^1 in $M(1;0)$), by lemma 1.10.10, X is not majorizable.

1.10.12   Theorem: Non-interpretability of $(t - ext)$ and (E_2).

The axiom of extensionality $(t - ext)$ and its special case (E_2) are neither $\wedge$- nor D-interpretable by primitive recursive functionals or by functionals from I-T.

Proof. By 0.1.14.4, $(t - ext)$ implies (E_2) already in pure logic so that a functional interpretation of $(t - ext)$ would yield a functional interpretation of (E_2). As (E_2) is a formula of type o, $(E_2)^\wedge$ and $(E_2)^D$ are equivalent by proposition 1.3.11. It therefore suffices to show that (E_2) is not $\wedge$-interpretable in I-T.
By theorem 1.10.8, functionals from I-T are majorizable, so, by theorem 1.10.11, they cannot $\wedge$-interpret (E_2).
The respective functional interpretation theorems now imply underivability of uninterpretable formulae:

1.10.13   Corollary: Underivability of $(t - ext)$ and (E_2)

The axiom of extensionality $(t-ext)$, even its special case (E_2), is underivable in theories $\wedge$- or D- interpretable in $T_\wedge$, $T_\wedge^c$ or in I-T, as there are

$$I\text{-}HA^{\omega+} \quad and \quad I\text{-}PA^{\omega*}$$

This follows from the last theorem by theorems 1.3.9 and 1.9.10.

Chapter 2

Analysis

Bar induction was introduced already by L. E. J. Brouwer in [5] as a constructively justifiable schema in intuitionistic analysis. Bar recursion in finite types was employed by Spector [38] for a Dialectica interpretation of classical analysis, essentially a weakly extensional version of PA_0^ω, extended by an appropriate classical choice principle. Howard [23] refines Spector's result considerably by first giving a Dialectica interpretation of bar induction by bar recursion and then proving this classical choice principle, formulated in the negative fragment, by bar induction, thus imbedding classical analysis into intuitionistic analysis. In Diller-Vogel [18], Howard's results are extended and generalized to intuitionistic analysis in finite types, independent of extensionality principles, replacing the Dialectica- by the ∧-translation. In the first three sections of this chapter, we follow this presentation. In section 2.4, strong computability of bar recursive functionals is shown following Vogel [50].

2.1 Bar recursive functionals

Bar recursion is usually introduced in (at least weakly) extensional systems (cf. Spector [38], Howard [23], Troelstra [45]). For a functional interpretation of intuitionistic analysis with a language in finite types, a combinatorial or neutral formulation of bar recursion is required. Such a formulation will be studied here. It uses the *restriction functional*

$$|_\tau := \lambda w^{o\to\tau}.R_{o\to\tau}(\lambda x^o u^{o\to\tau} y^o.D_\tau(uy)(wx)(x \dotminus y))0^{o\to\tau}$$

defined under 1.4.1, with D_τ the case distinction functional introduced in 1.1.5, together with its basic properties listed in lemma 1.4.2. We again write $c|t$ for $|_\tau ct$ and use the tuple notation of 0.1.13.

131

Moreover, such a combinatorial formulation of bar recursion also uses a one-point *concatenation functional* as follows:

2.1.1   Definition of concatenation $*_\tau$

For any type τ, a **concatenation functional** $*_\tau$ is defined as

$$*_\tau := \lambda cxuy.D_\tau((c|x)y)u(x \dot- y)$$

We write $c *_t u$ for $*_\tau ctu$ and thus have $(c *_t u)s = D((c|t)s)u(t \dot- s)$.

2.1.2   Lemma: Elementary properties of $*_\tau$

Concatenation $*_\tau$ satisfies provably in T:

$$
\begin{aligned}
&(1) \quad s < t \to (c *_t u)s = cs \\
&(2) \quad t \le s \to (c *_t u)s = u \\
&(3) \quad s \le t \to (c *_t u)|s = c|s \\
&(4) \quad (c *_t u)|suc\, t = c *_t u = (c|t) *_t u \\
&(5) \quad c *_t (ct) = c|suc\, t
\end{aligned}
$$

Proof. Since $s < t \to t \dot- s \ne 0$, by 1.1.5 and lemma 1.4.2 (3),

$$s < t \to (c *_t u)s = (c|t)s = cs$$

which is (1), and since $t \le s \to t \dot- s = 0$, (2) follows.
(3) is proved by induction on s.
1. $(c *_t u)|0 = 0^{o \to \tau} = c|0$.
2. As $suc\, x \le t \to x \le t \wedge x < t$, $suc\, x \le t$ implies by I.H. $(c *_t u)|x = c|x$ and therefore by (ξ') (cf. 0.1.11) also

$$
\begin{aligned}
(c *_t u)|(suc\, x) &= \lambda y.D(((c *_t u)|x)y)((c *_t u)x)(x \dot- y) \\
&= \lambda y.D((c|x)y)(cx)(x \dot- y) \qquad\qquad = c|suc\, x
\end{aligned}
$$

and (3) follows by (IND).
(4) is a consequence of (3), applying the induction step of (3) to $x = t \le t$ as follows:

$$
\begin{aligned}
(c *_t u)|(suc\, t) &= \lambda y.D(((c *_t u)|t)y)((c *_t u)t)(t \dot- y) \\
&= \lambda y.D((c|t)y)u(t \dot- y) \qquad\qquad = c *_t u = (c|t) *_t u
\end{aligned}
$$

The last equation holds, because $(c|t)|t = c|t$.
(5) Directly from the definitions of $|$ and $*_\tau$, we get

$$
\begin{aligned}
c *_t (ct) &= \lambda y.D(((c *_t (ct))|t)y)((c *_t (ct))t)(t \dot- y) \\
&= \lambda y.D((c|t)y)(ct)(t \dot- y) \qquad\qquad = c|suc\, t
\end{aligned}
$$

2.1.3 Stretching the tuple notation again

For concatenation, we use the same tuple notation as for restriction in 1.4.4: If τ is a type tuple $\tau_1, ..., \tau_n$ and if $c = c_1, ..., c_n$ and $u = u_1, ..., u_n$ are term tuples of type tuples $o \to \tau$ and τ respectively, $t : o$, then $c *_t u$ stands for the term tuple $c_1 *_t u_1, ..., c_n *_t u_n$. An equation between such tuples stands as before for the conjunction of the equations between corresponding elements of the tuples. Lemma 2.1.2 clearly holds also for such tuples c, u.

2.1.4 Background: Trees, nodes, paths, secured/unsecured nodes, well-founded subtrees

For any type or type tuple σ, $\mathbb{B}_\sigma$ denotes the **tree** of all finite sequences or **nodes**

$$\langle c_0, ..., c_{k-1} \rangle \qquad (k \in \mathbb{N})$$

of objects $c_i : \sigma$, with k the **length** of the node. A **path** in the tree $\mathbb{B}_\sigma$ is a function (or tuple of functions) $c : o \to \sigma$ which runs exactly through the nodes

$$\langle c_0, ..., c_{t-1} \rangle \qquad (t : o)$$

For this finite sequence of length t, the pair $c|t, t$ is used as a code, whereas $c|t$ alone designates an (ultimately constant) function of type $o \to \tau$ out of which the length term t cannot be extracted. In fact, if $Y : (o \to \sigma) \to o$ is a continuous functional (cf. definition 1.5.27 and the continuity theorem 1.5.28), for any $c : o \to \sigma$, Y can only exploit finitely many values ci of c to compute the value Yc. i.e., for sufficiently large $t \geq t_1$, $Y(c|t)$ remains constant. Taking $t_0 = 1 + max\, t_1\, (Y(c|t_1))$, we get $Y(c|t) < t$ for all $t \geq t_0$. In that case, the nodes $(c|t, t)$ are called **secured** by Y or Y-**secure**. For continuous Y, the nodes **unsecured** by Y (or Y-**unsecure**) therefore form a well-founded subtree of $\mathbb{B}_\sigma$. The assumed continuity of Y is thus the intuitive reason that the following schema of bar recursion defines total functionals.

Making use of the functionals $|_\tau$ and $*_\tau$, *bar recursion* and *bar induction* are introduced here without recourse to types for finite sequences of objects of a given type:

2.1.5 Bar recursive functionals $\mathbf{B}_{\sigma\tau}$ and bar recursion $Sim\mathbf{BR}_\sigma$

Let σ be a type tuple. For any type tuple τ, a **bar recursive functional** $\mathbf{B}_{\sigma\tau}$ of type tuple

$$((o \to \sigma) \to o) \to ((o \to \sigma) \to o \to \tau) \to ((\sigma \to \tau) \to (o \to \sigma) \to o \to \tau) \to (o \to \sigma) \to o \to \sigma \to \tau$$

is added to the constants of $T_\wedge$, and for any term touples Y, G, H, c, t, u of fitting type touples, the following **axioms of simultaneous bar recursion** - in which we make ample use of the abbreviation $t^+ := suc\,t$ - are added to the theory $T_\wedge$:

$$Y(c *_t u) < t^+ \to \mathsf{B}_{\sigma\tau} Y G H ctu =_\tau G(c *_t u)t^+$$

$SimBR_\sigma$

$$Y(c *_t u) \geq t^+ \to \mathsf{B}_{\sigma\tau} Y G H ctu =_\tau H(\mathsf{B}_{\sigma\tau} Y G H(c *_t u)t^+)(c *_t u)t^+$$

The resulting theory is denoted $T_\wedge + SimBR_\sigma$.

For a type tuple $\tau = \tau_1, ..., \tau_n$, the schema $SimBR_\sigma$ consists of $2n$ implications. It may be compressed to a schema of n equations as follows:

$$\mathsf{B}_{\sigma\tau} Y G H ctu = D(H(\mathsf{B}_{\sigma\tau} Y G H(c*_t u)t^+)(c*_t u)t^+)(G(c*_t u)t^+)(Y(c*_t u)\dot- t)$$

Proof. By definition 1.1.6, $Y(c *_t u) \leq t \leftrightarrow Y(c *_t u) \dot- t = 0$. Therefore, by definition and lemma 1.1.5 introducing the case distinction functional D, the alternative notation of bar recursion implies $SimBR_\sigma$. With proof by cases 1.1.4, also the other direction follows.

Making the dependance on $Y(c *_t u) \dot- t$ explicit as an argument of $\mathsf{B}_{\sigma\tau}$, a *companion* $\mathsf{B}^*_{\sigma\tau}$ of $\mathsf{B}_{\sigma\tau}$ is introduced by

2.1.6   Definition of the companion $\boldsymbol{\mathsf{B}^*_{\sigma\tau}}$ of $\boldsymbol{\mathsf{B}_{\sigma\tau}}$

$$\mathsf{B}^*_{\sigma\tau} Y G H ctxu = D(H(\mathsf{B}_{\sigma\tau} Y G H(c *_t u)t^+)(c *_t u)t^+)(G(c *_t u)t^+)x$$

Then $\mathsf{B}_{\sigma\tau}$ and $\mathsf{B}^*_{\sigma\tau}$ are connected as follows:

(1) $\mathsf{B}_{\sigma\tau} Y G H ctu = \mathsf{B}^*_{\sigma\tau} Y G H ct(Y(c *_t u) \dot- t)u$

(2) $\mathsf{B}^*_{\sigma\tau} Y G H ct0u = G(c *_t u)t^+$

(3) $\mathsf{B}^*_{\sigma\tau} Y G H cts^+u = H(\mathsf{B}_{\sigma\tau} Y G H(c *_t u)t^+)(c *_t u)t^+$

Equations (1), (2), (3) clearly follow from the above argument.

This variant will be used in section 2.4.

Setting $F := \mathsf{B}_{\sigma\tau} Y G H$, lemma 2.1.2 (4) implies

(∗) $Fctu = F(c|t)tu$

It is therefore no restriction, if in $SimBR_\sigma$ we put $u = ct$. Due to lemma 2.1.2 (5), the schema $SimBR_\sigma$ is then simplified to

$$Y(c|t^+) < t^+ \to \mathsf{B}_{\sigma\tau} Y G H ct(ct) = G(c|t^+)t^+$$

$SimBR_\sigma$

$$Y(c|t^+) \geq t^+ \to \mathsf{B}_{\sigma\tau} Y G H ct(ct) = H(\mathsf{B}_{\sigma\tau} Y G H ct^+)(c|t^+)t^+$$

2.1.7 Short form of bar recursion

For $F = \mathsf{B}_{\sigma\tau}YGH$, the term tupel F' defined by

$$F'cx = D(G(c|x)x)(H(Fcx)(c|x)x)(x \overset{-}{-} Y(c|x))$$

is called the **short form** of F. Clearly, F' also satisfies $(*)$.

2.1.8 Lemma

The short form F' of $F = \mathsf{B}_{\sigma\tau}YGH$ and F are related as follows:
(1') $Fctu = F'(c *_t u)t^+$
(2') $Y(c|t) < t \to F'ct = G(c|t)t$
(3') $Y(c|t) \geq t \to F'ct = H(Fct)(c|t)t$
The *proof* follows the pattern of the proof in 2.1.5 above.

2.1.9 Remark: Bar recursion under type-extensionality

The combinatorial form in which bar recursion is presented here is equivalent to the one given in Howard [24] and apparently weaker than the one presented in Spector [38], Howard [23], and Luckhardt [30] which runs as follows: For any type-tuple τ, there exists a tuple of bar functionals $\mathsf{B}_{\sigma\tau}^{\lambda}$ such that for $F' := \mathsf{B}_{\sigma\tau}^{\lambda}YGH$, the following axioms hold:

$$Y(c|t) < t \to F'ct = G(c|t)t$$
$$\lambda - SimBR_{\sigma}$$
$$Y(c|t) \geq t \to F'ct = H(\lambda u.F'(c *_t u)t^+)(c|t)t$$

This form of bar recursion is connected to $SimBR_{\sigma}$ via the rule of type-extensionality $(T - EXT)$ introduced in 0.1.14 as follows:

Lemma
On the basis of $T_{\wedge} + (T - EXT)$, the schemata $SimBR_{\sigma}$ and $\lambda - SimBR_{\sigma}$ are equivalent:

$$T_{\wedge} + (T - EXT) + SimBR_{\sigma} \quad \equiv \quad T_{\wedge} + (T - EXT) + \lambda - SimBR_{\sigma}$$

In fact, lemma 2.1.8 implies on the basis of $T_{\wedge} + (T - EXT)$:
(i) If $F = \mathsf{B}_{\sigma\tau}YGH$, then the short form F' of F satisfies $\lambda - SimBR_{\sigma}$, i.e. $F' = \mathsf{B}_{\sigma\tau}^{\lambda}YGH$.
(ii) If $F' := \mathsf{B}_{\sigma\tau}^{\lambda}YGH$, then F' is the short form of $F = \mathsf{B}_{\sigma\tau}YGH$ which satisfies $SimBR_{\sigma}$.
Apparently, the equivalence of the two schemata cannot be shown without recourse to the rule of type-extensionality $(T - EXT)$.

2.1.10　Remark: Bar recursion and product types

In the presence of product types and pairing functionals, the schema $SimBR_\sigma$ may in $T_\wedge$ be derived from the schema BR of simple bar recursion in which the type tuples σ and τ consist of one type each. Whereas simultaneous primitive recursion $(SimR)$ is derivable from the schema (R) of (simple) primitive recursion in Gödel's theory T in linear types, as shown in section 1.4, it remains an open problem whether analogously the schema $SimBR_\sigma$ can be derived in the theory $T_\wedge + BR$ in linear types.

2.1.11　Remark on bounded quantifiers in type o context

Bounded quantifiers in front of formulae of type o are definable already in T_0 and have characteristic terms as discussed after proposition 1.1.15. In particular, be it provably in T_0 or just as a convenient notation, we have the equivalences (where x may occur in s, but not in t)

$$\forall x \le t \; s \ge x \Leftrightarrow \forall x < t^+ \; s \ge x \Leftrightarrow max\{x \dot{-} s \mid x < t^+\} = 0$$
$$\exists x \le t \; s < x \Leftrightarrow \neg\forall x < t^+ \; s \ge x \Leftrightarrow max\{x \dot{-} s \mid x < t^+\} > 0$$

Bar recursion is frequently stated in the form of *monotone bar recursion* which uses these bounded quantifiers:

2.1.12　Proposition: Monotone bar recursion

Let σ, τ be type tuples. Given a term $Y : (o \to \sigma) \to o$ and tuples of terms G, H of fitting types, a tuple F of terms is definable in $T_\wedge + SimBR_\sigma$ satisfying the schema of **monotone bar recursion** :

$$MonBR_\sigma \quad \begin{aligned} &\exists x \le t^+ \; Y(c|x) < x \to \; Fct(ct) = G(c|t^+)t^+ \\ &\forall x \le t^+ \; Y(c|x) \ge x \to \; Fct(ct) = H(F(c|t^+)t^+)(c|t^+)t^+ \end{aligned}$$

Proof. Given Y, G, H as stated, define a tuple H_1 of terms by

$$H_1 wct = D(Gct)(Hwct)(max\{x \dot{-} Y(c|x) \mid x < t\})$$

and put $F := \mathsf{B}_{\sigma\tau} Y G H_1$. Since already in $T_\wedge$

$$max\{x \dot{-} Y(c|x) \mid x < t^+\} = 0 \leftrightarrow \forall x < t^+ \; x \le Y(c|x),$$

we have by lemma 1.1.5
(1)　$\exists x < t^+ \; Y(c|x) < x \to H_1 wct^+ = Gct^+$
(2)　$\forall x < t^+ \; Y(c|x) \ge x \to H_1 wct^+ = Hwct^+$

The proposition now follows by considering the following cases:

Case 1. $Y(c|t^+) < t^+$. Then by $SimBR_\sigma$,

$Fct(ct) = G(c|t^+)t^+$

Case 2. $Y(c|t^+) \geq t^+$. Then again by $SimBR_\sigma$,

$Fct(ct) = H_1(F(c|t^+)t^+)(c|t^+)t^+$

Subcase 2.1. In addition, $\exists x < t^+ \; Y(c|x) < x$. Then by (1)

$Fct(ct) = G(c|t^+)t^+$

Subcase 2.2. Alternatively, $\forall x < t^+ \; Y(c|x) \geq x$. Then by (2)

$Fct(ct) = H(F(c|t^+)t^+)(c|t^+)t^+$

Case 1 and subcase 2.1 yield the first line, case 2 and subcase 2.2 yield the second line of $MonBR_\sigma$.

2.1.13 Background continued: Fans in $\mathbb{B}_0$ and $\mathbb{B}_\sigma$, width of a fan

The **tree** $\mathbb{B}_0$, as the special case $\sigma = o$ of background 2.1.4 above, consists of all finite sequences or nodes $\langle n_0, ..., n_{k-1} \rangle$ of natural numbers $n_i, i < k$. **Paths** in that tree are choice sequences, i.e. functions $w : o \to o =: 1$. A path w **runs through** the nodes $w|k, k$, a pair of terms of type tuple $1, o$. Whereas nodes of $\mathbb{B}_0$ may, of course, be coded by type o objects, it makes sense to stay with the kind of coding chosen for nodes in the general case. Any function $b : \mathbb{B}_0 \to \mathbb{N}$, i.e. $b : 1 \to 1$, cuts out a **fan** $\mathbb{F}_0(b)$ out of $\mathbb{B}_0$, containing exactly those paths w for which

(*) $\qquad\qquad wk < b(w|k)k$

holds for all $k \in \mathbb{N}$; so a node $w|n, n$ belongs to $\mathbb{F}_0(b)$ iff (*) holds for all $k < n$. In node $w|n, n$, the fan $\mathbb{F}_0(b)$ has **width** $b(w|n)n$.

Any function $g : \mathbb{B}_0 \to \sigma$ induces a map

$g^{<>} : \mathbb{B}_0 \to \mathbb{B}_\sigma \qquad g^{<>} : (w|k), k \longmapsto \langle g(w|1)0, ...g(w|k)(k-1) \rangle$

This way, g enumerates a subtree $g(\mathbb{B}_0)$ of $\mathbb{B}_\sigma$. g, restricted to the fan $\mathbb{F}_0(b)$ in $\mathbb{B}_0$ (with b satisfying (*)) enumerates the subfan $g(\mathbb{F}_0(b))$ of $g(\mathbb{B}_0)$ as well as of $\mathbb{B}_\sigma$. For the functions b and g, both defined on $\mathbb{B}_0$, it means no restriction if we assume for all $w : 1$ and $x : o$

(**) $\qquad\qquad gw|x = g(w|x)|x$ and $bwx = b(w|x)x$

In that case, gw is a path in $\mathbb{B}_\sigma$ running exactly through the nodes $g^{<>}(w|k, k)$, coded by $gw|k, k$.

We collect some information about fans in $\mathbb{B}_0$ as well as in $\mathbb{B}_\sigma$ in general.

For any function $f : \mathbb{B}_0 \to \mathbb{N}$, the maximum of all values of f at nodes in the fan $\mathbb{F}_0(b)$ of a fixed length t is computed as follows:

2.1.14   Lemma: Fan-maximum

There is a primitive recursive functional MAX *which, given* $b : 1 \to 1$ *cutting out the fan* $\mathbb{F}_0(b)$ *out of* $\mathbb{B}_0$, *and given* $f : 1 \to 1$, *computes, for any* $t : o$, *the maximum of the values of* f *at the nodes in* $\mathbb{F}_0(b)$ *of length* t, *i.e.*

(1) $MAXbft = max\{f(w|t)t \mid \forall x < t \; wx < b(w|x)x\}$

Proof. Define H and MAX in $T_\wedge$ by

$$Hbfwt = max\{f(w *_t y)t^+ \mid y < b(w|t)t\} \text{ and } MAXbft = J(Hb)ft0^10$$

with J the iterator introduced in 1.1.5. We then have

(2) $MAXbf0 \quad = f0^10 \qquad$ and, using (1) as I.H.,

(3) $MAXbft^+ = MAXb(Hbf)t$

$\qquad = max\{max\{f(w *_t y)t^+ \mid y < b(w|t)t\} \mid \forall x < t \; wx < b(w|x)x\}$

$\qquad = max\{f(w|t^+)(t^+ \mid \forall x < t^+ \; wx < b(w|x)x\}$

The lemma then follows by (IND) from (2) and (1) $\to$ (3).

2.1.15   Definition: Unsecured part of a fan

Let a tuple of terms g, b, Y of type tuple $1 \to o \to \sigma, 1 \to 1, (o \to \sigma) \to o$ be given and assume (**) as in 2.1.13. Then

$$UFAN[w, t] :\equiv \forall x < t(wx < bwx \wedge Y(gw|x) \geq x)$$

expresses that, up to t, the path gw runs within the fan $g(\mathbb{F}_0(b))$ through nodes unsecured by Y.

2.1.16   Lemma

$$T_\wedge \vdash UFAN[w, t] \wedge Y(gw|t) \geq t \to \forall y < bwt \; UFAN[w *_t y, t^+]$$

Proof. Clearly, in $T_\wedge$

$\forall x < t \; wx < bwx \to \forall y < bwt \forall x < t^+ \; (w *_t y)x < b(w *_t y)t$

and since $g(w *_t y)|t = gw|t$, also

$\forall x < t \; Y(gw|x) \geq x \wedge Y(gw|t) \geq t \to \forall x < t^+ \; Y(gw|x) \geq x$

The conjunction of these two implications yields the lemma.

Howard [23], lemma 3C, applies Kreisel's idea [29] to single paths in the tree $\mathbb{B}_\sigma$. The following proposition transfers this idea to fans in $\mathbb{B}_\sigma$.

2.1.17 Proposition: Estimating the depth of the unsecured part of a fan

Let g, b, Y be as above in definition 2.1.15, and assume (**) as in 2.1.13. Then there is a term L, definable in $T_\wedge + SimBR_\sigma$, such that

(1) $\exists x \le t\ Y(c|x) < x \to Lct = 0$,

(2) $\forall x \le t\ Y(c|x) \ge x \to Lct = 1 + max\{L(c *_t gwt)t^+ \mid \forall x \le t\ wx < bwx\}$

and for $t_0 := L0^{o \to \sigma}0$,

(3) $T_\wedge + SimBR_\sigma \vdash UFAN[w, t_0] \to Y(gw|t_0) < t_0$:

The subfan enumerated in $\mathbb{B}_\sigma$ by g, b and wellfounded by Y has depth $\le t_0$.

Proof. By $MonBR_\sigma$, define a term F with $Fctu = F(c|t)tu$ and

$$\exists x \le t^+\ Y(c|x) < x \to Fct(ct) = 0$$
$$\forall x \le t^+\ Y(c|x) \ge x \to Fct(ct) =$$
$$1 + MAX\, b\big(\lambda wx.F(c|t^+)t^+(gw(prd\,x))\big)(suc\,t^+).$$

Let L be the short form of F. Then $L(c|t)t = Lct$ by 2.1.7; by 2.1.8, we have (1) and

$$\forall x \le t^+\ Y(c|x) \ge x \to$$
$$Lct\quad = 1 + MAX\, b\big(\lambda wx.F(c|t)t(gw(prd\,x))\big)(t^+)$$
$$= 1 + max\{F(c|t)t(g(w|t^+)t) \mid \forall x < t^+\ wx < bwx\}$$
$$= 1 + max\{L(c *_t (gwt))t^+ \mid \forall x < t^+\ wx < bwx\}$$

by (**), and that is (2). (1) and (2) imply

$\forall x < t^+\ wx < bwx \to L(c *_t (gwt))t^+ \le prd\,(Lct)$, in particular for $c = gw$

$\forall x < t^+\ wx < bwx \to L(gw)t^+ \le prd\,(L(gw)t)$.

By iterating this last step, i.e. by induction on s, we obtain

$\forall x < t + s\ wx < bwx \to L(gw)(t + s) \le L(gw)t \dot- s$.

For $t = 0$ and $s = t_0 = L(w|0)0$, this implies

$\forall x < t_0\ wx < bwx \to L(gw)t_0 \le L(gw)0 \dot- t_0 = 0$.

Now assume $UFAN[w, t_0]$. Then, on the one hand, $\forall x < t_0\ wx < bwx$ which implies $L(gw)t_0 = 0$, as just shown, and hence $\exists x \le t_0\ Y(gw|x) < x$ in contraposition to (2). On the other hand, $\forall x < t_0\ Y(gw|x) \ge x$. That amounts to $Y(gw|t_0) < t_0$, as was to be shown.

2.1.18 Remark on the exact depth of the unsecured part of a fan

The term t_0 in (3) is in general larger than the depth of the by Y unsecured part of the fan $g(\mathbb{F}_0(b))$ enumerated in $\mathbb{B}_\sigma$ by g. An exact expression for the depth of the Y-unsecured part of $g(\mathbb{F}_0(b))$ is obtained as follows in

$T_\wedge + SimBR_\sigma + BR_o$:

We first compute the length of the unsecured part of any single path running through the subtree $g(\mathbb{B}_o)$ of $\mathbb{B}_\sigma$ by defining a term L by means of $MonBR_\sigma$ such that

$$\exists x \leq t\, Y(c|x) < x \to Lwct = 0$$
$$\forall x \leq t\, Y(c|x) \geq x \to Lwct = 1 + Lw(c *_t gwt)t^+$$

$L_1 w := Lw0^{o \to \sigma}0$ is the length of the Y-unsecured part of the path gw in $\mathbb{B}_\sigma$, as is seen by IND. Then we compute the maximum of the lengths $L_1 w$ with the paths w running through the subfan $\mathbb{F}_o(b)$ of $\mathbb{B}_o$. That is, we define by BR_o a term L_0 such that

$$L_1(w|t) < t \to L_0 wt = 0$$
$$L_1(w|t) \geq t \to L_0 wt = 1 + max\{L_0(w *_t y)t^+ \mid y < bwt\}$$

$t_0 := L_0 0^1 0$ is now the exact depth of the Y-unsecured part ot the fan $g(\mathbb{F}_0(b))$, satisfying in particular (3) above.

The proof of proposition 2.1.17 given above in detail avoids application of BR_o and replaces it in this specific case by the primitive recursive functional MAX.

The generalized induction rule proved in proposition 1.1.22 is admissible in $T_\wedge$ and by an application of the substitution rule also directly derivable. Below, we apply it as an inference "from $suc\,x$ to x":

2.1.19   Corollary to proposition 1.1.22

In $T_\wedge$, augmented by the substitution rule, the following rule of "induction downwards" is directly derivable:

$$A[t, z],\ \forall y < s_1[x]A[suc\,x, a_1[x]] \to A[x, z] \vdash A[0, z],$$

where a_1 and z are tuples of terms and variables resp. of the same type tuple, and y does not occur in s_1, A, t and x not in t.

Proof. The direction of induction in 1.1.22 is inverted by substituting $t \dot- suc\,x$ for x and putting
$s[x] := s_1[t \dot- suc\,x]$ and $a[x] := a_1[t \dot- suc\,x]$.
The premises then imply

$$0 \leq t \to A[t \dot- 0, z]$$
$$\forall y < s[x](x \leq t \to A[t \dot- x, a[x]]) \to (suc\,x \leq t \to A[t \dot- suc\,x, z])$$

For $B[x, z] :\equiv x \leq t \to A[t \dot- x, z]$, these are literally the premises of proposition 1.1.22, by which now the derivability in $T_\wedge$ of $B[t, z]$ and equivalently of $A[0, z]$ follows.

From proposition 2.1.17 and this corollary we conclude:

2.1.20   Theorem: Generalized fan rule

In $T_\wedge + SimBR_\sigma$, augmented by the substitution rule, the following generalized fan rule is directly derivable:

$$Y(c|x) < x \to D[c|x, x, z]$$

$$\vdash D[0^{o \to \sigma}, 0, z]$$

$$\forall y < X(c|x)xz \; D[c *_x U(c|x)xyz, x^+, V(c|x)xyz] \to D[c|x, x, z]$$

where Y, X, U, V form a term tuple of suitable type tuple in which the variables of the tuple c, x, y, z of type tuple $o \to \sigma, o, o, \tau$ do not occur, with y also not occurring in D.

Proof. By simultaneous course-of-values recursion, theorem 1.4.4, a term tuple g, h is defined by

$$gwt \;\; = U(gw|t)t(wt)(hwt)$$
$$hw0 \;\; = z$$
$$hwt^+ = V(gw|t)t(wt)(hwt)$$

Moreover, a term b is explicitly defined by $bwt = X(gw|t)t(hwt)$.

By induction on x, $(**)$ in 2.1.13 and $hwx = h(w|x)x$ follow. In the premises of the theorem, we substitute gw for c and hwx for z and obtain

(1) $Y(gw|x) < x \to D[gw|x, x, hwx]$

(2) $\forall y < bwx \; D[g(w *_x y)|x^+, x^+, h(w *_x y)x^+] \to D[gw|x, x, hwx]$

If in (1) we replace x by the term $t_0 := L0^{o \to \sigma}0$ defined in proposition 2.1.17 in terms of g, b, Y, (3) in that proposition implies

(3) $UFAN[w, t_0] \to D[gw|t_0, t_0, hwt_0]$

Lemma 2.1.16 implies

(4) $Y(gw|x) \geq x \to (\forall y < bwx \; UFAN[w *_x y, x^+] \to D[gw|x, x, hwx])$
$\to (UFAN[w, x] \to D[gw|x, x, hwx])$

Because of (1), the antecedent $Y(gw|x) \geq x$ in (4) is redundant. So (4) and (2) imply

(5) $\forall y < bwx \; (UFAN[w *_x y, x^+] \to D[g(w *_x y)|x^+, x^+, h(w *_x y)x^+])$
$\to (UFAN[w, x] \to D[gw|x, x, hwx])$

Now we apply corollary 2.1.19 to (3) and (5) and obtain

$$UFAN[w, 0] \to D[gw|0, 0, hw0]$$

which immediately yields $D[0^{o \to \sigma}, 0, z]$, and the theorem is proved.

This generalized fan rule plays the same rôle in the functional interpretation of bar induction (cf. next section) that the generalized induction rule plays in the functional interpretation of the induction rule (IND) in section 1.2. It should be noticed that the proof of the generalized fan rule in $T_\wedge + SimBR_\sigma$ does not require any special form of bar induction, whereas the proof of the generalized induction rule in $T_\wedge$ certainly requires applications of (IND).

2.2   Intuitionistic analysis and its functional interpretations

Intuitionistic analysis in finite types is here introduced in the spirit of L.E.J. Brouwer as an extension of HA^ω, even of $HA^{\omega+}$ (cf. 1.2.11), by a schema of bar induction, also in finite types.

2.2.1   *Bar induction in finite types*

Let σ be a type tuple. (Simultaneous) **bar induction of type tuple** σ is the schema

$$BI_\sigma \qquad Hyp\,1 \wedge Hyp\,2 \wedge Hyp\,3 \wedge Hyp\,4 \rightarrow Q[0^{o\rightarrow\sigma}, 0]$$

where $P[c, x], Q[c, x]$ are formulae of $L(HA^\omega)$ and

$$
\begin{aligned}
Hyp\,1 \qquad & \forall c^{o\rightarrow\sigma} \exists y\, P[c|y, y] \\
Hyp\,2 \qquad & \forall c \forall x \forall y < x (P[c|y, y] \rightarrow P[c|x, x]) \\
Hyp\,3 \qquad & \forall c \forall x (P[c|x, x] \rightarrow Q[c|x, x]) \\
Hyp\,4 \qquad & \forall c \forall x (\forall u\, Q[c *_x u, suc\, x] \rightarrow Q[c|x, x]).
\end{aligned}
$$

The theory $HA^{\omega+} + BI_\sigma$ is also called **intensional intuitionistic analysis of type** σ.

$Hyp\,1$ expresses the well-foundedness of the part of $\mathbb{B}_\sigma$ unsecured by P.
$Hyp\,2$ is the upward monotonicity of P.
$Hyp\,3$ is the transfer from P to Q, $P \subset Q$.
$Hyp\,4$ is the bar induction step for Q.
The conclusion $Q[0^{o\rightarrow\sigma}, 0]$, expressing that Q applies to the root of $\mathbb{B}_\sigma$, in conjunction with $Hyp\,1$ to $Hyp\,4$ means in fact that Q applies to all nodes of $\mathbb{B}_\sigma$.

The central aim of this section is a proof of:

2.2.2   *Theorem: $\wedge$-interpretation of intuitionistic analysis*

Intensional intuitionistic analysis of type σ is $\wedge$-interpretable in $T_\wedge + SimBR_\sigma$.

$$HA^{\omega+} + BI_\sigma \overset{\wedge}{\hookrightarrow} T_\wedge + SimBR_\sigma$$

Since the qf axioms $SimBR_\sigma$ are $\wedge$-translated identically, by the extended $\wedge$-interpretation theorem 1.2.13 (2) we already have

$$HA^{\omega+} + SimBR_\sigma \overset{\wedge}{\hookrightarrow} T_\wedge + SimBR_\sigma$$

It therefore suffices to prove

2.2.3 Theorem

$$HA^{\omega+} + SimBR_\sigma \vdash BI_\sigma$$

Before this theorem is proved, a couple of lemmata should be discussed.

2.2.4 Lemma

Let $A[y], B[z]$ be two prenex, purely universal formulae in $L(HA^\omega)$ with disjoint tuples of variables y, z. Then

$$HA^\omega + AC + IP_\wedge \vdash (\exists y A[y] \to \exists z B[z]) \to \exists Z \forall y (A[y] \to B[Zy])$$

Proof. $(\exists y A[y] \to \exists z B[z])$ is logically equivalent to $\forall y (A[y] \to \exists z B[z])$ which by $IP_\wedge$ implies $\forall y \exists z (A[y] \to B[z])$ which by AC implies $\exists Z \forall y (A[y] \to B[Zy])$.

In analogy to Howard [23], we can now partially translate the hypotheses of bar induction:

2.2.5 Lemma: Partial $\wedge$-translation of bar induction

Let $P[c, x]^\wedge \equiv \exists r P'[c, x, r] \equiv \exists r \forall s P_\wedge[c, x, r, s]$
and $Q[c, x]^\wedge \equiv \exists v Q'[c, x, v] \equiv \exists v \forall z Q_\wedge[c, x, v, z]$
with pairwise disjoint tuples of variables c, x, r, s, v, z. Then for $i = 1, 2, 3, 4$

$$HA^{\omega+} \vdash Hyp\, i \leftrightarrow \exists Y\, RLSH(Hyp\, i)'$$

where $(Hyp\, i)'$ is the following 'partial $\wedge$-translation' of $Hyp\, i$:

$(Hyp\, 1)'$ $\forall c P'[c|(Yc), Yc, Rc]$
$(Hyp\, 2)'$ $\forall cxr \forall y < x(P'[c|y, y, r] \to P'[c|x, x, Lcxyr])$
$(Hyp\, 3)'$ $\forall cxr(P'[c|x, x, r] \to Q'[c|x, x, Scxr])$
$(Hyp\, 4)'$ $\forall cxw(\forall u Q'[c *_x u, suc\, x, wu] \to Q'[c|x, x, Hwcx])$

Proof. We have $(Hyp\, 1)^\wedge \equiv \exists Y R\, (Hyp\, 1)'$; so the equivalence for $i = 1$ holds in $HA^{\omega+}$.
As $HA^{\omega+} \vdash P[c, x] \leftrightarrow \exists r P'[c, x, r]$, we have in $HA^{\omega+}$
$Hyp\, 2 \leftrightarrow \forall cx \forall y < x(\exists r P'[c|y, y, r] \to \exists l\, P'[c|x, x, l]) \leftrightarrow \exists L\, (Hyp\, 2)'$

by lemma 2.2.4. Similarly, in $HA^{\omega+}$

$Hyp\ 3 \leftrightarrow \forall cx(\exists r P'[c|x, x, r] \to \exists s\, Q'[c|x, x, s]) \leftrightarrow \exists S\ (Hyp\ 3)'$,

again by lemma 2.2.4. Finally, again in $HA^{\omega+}$

$Hyp\ 4 \leftrightarrow \forall cx(\exists w \forall u Q'[c *_x u, suc\, x, wu] \to \exists v Q'[c|x, x, v]) \leftrightarrow \exists H(Hyp\ 4)'$,

also by lemma 2.2.4.

2.2.6   Definition

Let $(Hyp)'$ denote the conjunction of the formulae $(Hyp\ i)'$, $(i = 1, 2, 3, 4)$, and let Y, R, L, S, H be a tuple of variables (or terms) of type tuple as in lemma 2.2.5. Abbreviate $\lambda cx.Scx(Lcx(Yc)(Rc))$ by G and $\mathbf{B}_{\sigma\tau}YGH$ by F and let F' be the short form of F as in 2.1.7.

2.2.7   Lemma

In $HA^{\omega} + SimBR_{\sigma}$ is derivable

  (1) $(Hyp)'$ $\to Y(c|x) < x \to Q'[c|x, x, F'cx]$

  (2) $(Hyp)' \to \forall u Q'[c *_x u, suc\, x, F'(c *_x u)(suc\, x)] \to Q'[c|x, x, F'cx]$

Proof. Assume $Y(c|x) < x$. By 1.4.2 (4), $c|x|(Y(c|x)) = c|(Y(c|x))$. This implies by $(Hyp\ 1)'$

$P'[c|Y(c|x), Y(c|x), R(c|x)]$. From this we get by $(Hyp\ 2)'$

$P'[c|x, x, L(c|x)x(Y(c|x))(R(c|x))]$, and from this by $(Hyp\ 3)'$ finally

$Q'[c|x, x, S(c|x)x(L(c|x)x(Y(c|x))(R(c|x)))] \equiv Q'[c|x, x, G(c|x)x]$.

In view of lemma 2.1.8 (1), this implies (1). However, it implies (2) only under the additional assumption $Y(c|x) < x$.

Assume $Y(c|x) \geq x$. Then $(Hyp\ 4)'$ implies

$\forall u Q'[c *_x u, suc\, x, F(c|x)xu)] \to Q'[c|x, x, H(F(c|x)x)(c|x)x]$,

and by lemma 2.1.8 (2) and (3) we obtain our claim (2) also in this case. So the lemma is proved.

2.2.8   Proof of theorem 2.2.3

We use the notation of lemma 2.2.5 and definition 2.2.6 and put

$$D[c, x, z] :\equiv Q_{\wedge}[c|x, x, F'cx, z]$$

Then $D[c, x, z] \leftrightarrow D[c|x, x, z]$

by $(*)$ in 2.1.6, and the $\wedge$-translation of the formula

$\forall cx(Y(c|x) < x \to Q'[c|x, x, F'(c|x)x])$ runs

$(1^{\wedge})$ $\forall cxz(Y(c|x) < x \to D[c|x, x, z])$

Similarly, the $\wedge$-translation of

$\forall cx(\forall u Q'[c *_x u, sux\, x, F'(c *_x u)(suc\, x)) \rightarrow Q'[c|x, x, F'(c|x)x])$ is

$(2^\wedge)$ $\quad \exists XUV\forall cxz$

$\qquad \left(\forall y < X(c|x)xz\, D[c *_x U(c|x)xyz, x^+, V(c|x)xyz] \rightarrow D[c|x, x, z] \right).$

Now, $HA^\omega + SimBR_\sigma$ is an extension of $T_\wedge + SimBR_\sigma$ satisfying the deduction theorem. Therefore, the generalized fan rule of theorem 2.1.20 is derivable in this theory as a schema. Applied to our situation, that means

$$(1^\wedge) \wedge (2^\wedge) \rightarrow \forall z D[0^{o\rightarrow\sigma}, 0, z]$$

is derivable in $HA^\omega + SimBR_\sigma$. As $\forall z D[0, 0, z]$ is the formula $Q'[0, 0, F'00]$ which certainly implies $Q^\wedge[0, 0]$, we also have in $HA^\omega + SimBR_\sigma$

$$(1^\wedge) \wedge (2^\wedge) \rightarrow Q^\wedge[0^{o\rightarrow\sigma}, 0].$$

So, by lemma 2.2.5,

$$HA^{\omega+} + SimBR_\sigma \vdash (Hyp)' \rightarrow Q^\wedge[0, 0]$$

and therefore also

$$HA^{\omega+} + SimBR_\sigma \vdash \exists YRLSH\, (Hyp)' \rightarrow Q^\wedge[0, 0].$$

However, by lemma 2.2.5, the conjunction of the four hypothesis $Hyp\, i$ ($i = 1, 2, 3, 4$) of BI_σ implies $\exists YRLSH\, (Hyp)'$ in $HA^{\omega+}$, so that finally

$$HA^{\omega+} + SimBR_\sigma \vdash BI_\sigma.$$

That completes the proof of theorem 2.2.3 and, as mentioned above, also of theorem 2.2.2, the $\wedge$-interpretation of intuitionistic analysis $HA^{\omega+} + BI_\sigma$ in $T_\wedge + SimBR_\sigma$.

2.2.9 *Corollary*

$HA^{\omega+} + BI_\sigma + SimBR_\sigma$ *is a conservative extension of* $T_\wedge + SimBR_\sigma$.

This is clearly a consequence of theorem 2.2.2 and the fact that the $\wedge$-translation is identity on the language of $T_\wedge + SimBR_\sigma$.

Theorems 2.2.2 and 2.2.3 can be transferred, within certain limits, to the Dialectica interpretation. As in section 1.3, the Dialectica interpretation is correct only on type o fragments of the theories considered above.

2.2.10 *Definition of type o versions of $SimBR_\sigma$, BI_σ, and associated theories*

As a type o version of $SimBR_\sigma$, we may take the one-equation formulation of bar recursion in 2.1.5 under type o context. Writing again t^+ for $suc\, t$, that amounts to the schema

$$Y(c *_t u) < t^+ \to d[\mathsf{B}_{\sigma\tau}YGHctu] = d[G(c *_t u)t^+]$$
$$Y(c *_t u) \geq t^+ \to d[\mathsf{B}_{\sigma\tau}YGHctu] = d[H(\mathsf{B}_{\sigma\tau}YGH(c *_t u)t^+)(c *_t u)t^+]$$

with an arbitrary context $d[z^\tau] : o$.

The type o version of BI_σ is, of course, the schema BI_σ, restricted to type o formulae.

The theories $T_0 + SimBR_\sigma$ and $HA_0^{\omega+} + SimBR_\sigma$ are the theories T_0 and $HA_0^{\omega+}$ respectively, each extended by the type o version of $SimBR_\sigma$. $HA_0^{\omega+} + BI_\sigma$ is $HA_0^{\omega+}$, extended by the type o version of BI_σ.

With this terminology, results analogous to theorems 2.2.2 and 2.2.3 follow also for the Dialectica interpretation. Theorem 1.3.9 immediately implies

$$HA_0^{\omega+} + SimBR_\sigma \xrightarrow{\wedge} T_0 + SimBR_\sigma$$

Inspection of the proof of theorem 2.2.3 shows: A proof of the type o version of BI_σ can be carried out already in $HA_0^{\omega+} + SimBR_\sigma$. So we also have a proof of the results of Howard [23] who considers a weakly extensional version of $SimBR_\sigma$:

2.2.11 *Theorem: Dialectica-interpretation of intuitionistic analysis*

$$HA_0^{\omega+} + SimBR_\sigma \vdash BI_\sigma$$

Therefore, the type o version of intuitionistic analysis is D-interpretable in $T_0 + SimBR_\sigma$:

$$HA_0^{\omega+} + BI_\sigma \xrightarrow{D} T_0 + SimBR_\sigma$$

Non-reductive modified realization does not yield a reduction of bar induction to bar recursion like the $\wedge$- and D-interpretations in theorems 2.2.3 and 2.2.11, due to the fact that modified realization translation does not reduce an infinite well-founded tree to a finite fan or even a finite path. The presence of Markov's principle $M_\wedge$ in the schema $A \leftrightarrow A^\wedge$ is essential.

2.3   Classical analysis and its functional interpretations

In his ground-breaking paper [38], C. Spector gives a Dialectica-interpretation of classical analysis in the theory $T_0 + (T - EXT) + BR$. The system of classical analysis he considers is the theory $PA_0^{\omega*} + AC_{o\tau}^-$ (cf. section 1.9). Howard, in [23], instead considers, for any type tuple τ, the negative version of the axiom schema

$$(AC + DC)_\tau \quad \forall x^o X^\tau \exists Y^\tau A[x, X, Y] \to \exists F^{o\to\tau} \forall x^o A[x, Fx, F(suc\, x)]$$

which contains the schema $AC_{o\tau}$ as well as the schema of dependent choice

$$DC_\tau \quad \forall X^\tau \exists Y^\tau A[X, Y] \to \exists F^{o\to\tau} \forall x^o A[Fx, F(suc\, x)]$$

As is easily seen, the schema $(AC + DC)_\tau$ is in turn a special case of the following schema of **choice depending on courses of values**:

$$CDC_\tau \quad \forall x^o X^{o\to\tau} \exists Y^\tau A[x, X|x, Y] \to \exists F^{o\to\tau} \forall x^o A[x, F|x, Fx]$$

In theories like $HA^\omega + (T - EXT)$ which allow coding of finite sequences of objects of type τ into one object of type τ, $(AC + DC)_\tau$ is equivalent to CDC_τ.

While the choice principles as they are put down here in the full language of HA^ω have simple functional interpretations in primitive recursive functionals and hence a straight forward constructive meaning, their negative versions are harder to interpret constructively and call for bar functionals for their functional interpretation.

Howard in [23] extends Spector's Dialectica interpretation to the theory $PA_0^{\omega*} + (T - EXT) + (AC + DC)_\tau^-$. Our major aim here, following essentially Diller-Vogel [18], is a generalization of his result, a $\wedge$-interpretation of classical analysis in all finite types with classical choice principle CDC_τ^- as described above, by bar-recursive functionals:

2.3.1   Theorem: $\wedge$-interpretation of classical analysis

$$PA^{\omega+} + CDC_\tau^- \overset{\wedge}{\hookrightarrow} T_\wedge^c + SimBR_\tau$$

In view of the $\wedge$-interpretation of intuitionistic analysis, theorem 2.2.2, this theorem is implied by

2.3.2   Theorem

$$PA^{\omega+} + CDC_\tau^- \prec HA^{\omega+} + Stab(=) + BI_\tau$$

Since by proposition 1.9.8, $PA^{\omega+} \prec HA^{\omega+} + Stab(=)$, for a proof of this theorem, it suffices to prove the following theorem of intuitionistic analysis:

2.3.3   Theorem

$$HA^{\omega+} + BI_\tau \vdash CDC_\tau^-$$

A first step towards a proof of this theorem is

2.3.4   Proposition

For any formula $A[x^o, X^{o\to\tau}, Y^\tau]$ of $L(HA^\omega)$, the theory $HA^\omega + BI_\tau$ proves

$$\neg(\forall F^{o\to\tau}\exists y^o\neg\forall x < y\, A[x, F|x, Fx] \wedge \forall x^o X^{o\to\tau}\neg\neg\exists Y^\tau\, A[x, X|x, Y])$$

This proposition claims slightly more than Howard [23], appendix, theorem 1, and Howard-Kreisel [25], theorem (ii), in that both assumptions
(1) $\forall F^{o\to\tau}\exists y^o\neg\forall x < y\, A[x, F|x, Fx]$
(2) $\forall x^o X^{o\to\tau}\neg\neg\exists Y^\tau\, A[x, X|x, Y])$
- out of which a contradiction is to be derived - are weaker than in the sources quoted.

Proof of the proposition. Abbreviate $\forall x < y\, A[x, c|x, cx]$ by $B[c, y]$ and put

$$P[c, x] :\equiv Q[c, x] :\equiv \neg B[c, x]$$

Assumption (1) immediately implies $Hyp\,1$; $Hyp\,2$ and $Hyp\,3$ are obvious for these P and Q.
We come to $Hyp\,4$. Assume
(3) $\forall u\, Q[c *_y u, suc\,y]$
which by lemma 1.4.2 is equivalent to
(4) $\forall u\neg(B[c|y, y] \wedge A[y, c|y, u])$
which is logically equivalent to
(5) $\neg(B[c|y, y] \wedge \exists u\, A[y, c|y, u])$
Assumption (2) implies
(6) $\neg\neg\exists u\, A[y, c|y, u]$
(5) and (6) together imply $\neg B[c|y, y]$ by positive logic, i.e.
(7) $Q[c|y, y]$.
However, (3) $\to$ (7) is $Hyp\,4$. So, by BI_τ, we obtain

$$Q[0^{o\to\tau}, 0] \equiv \neg B[0^{o\to\tau}, 0] \equiv \neg\forall x < 0\, A[x, 0^{o\to\tau}|x, 0^{o\to\tau}x]$$

contradicting axiom $B1$. So the proposition is proved.

2.3.5 Lemma

For any quantifier-free formula $B[y, w]$

$$HA^\omega + M_\wedge \vdash \neg\forall y^o w\, B[y, w] \to \exists y \neg\forall x < y\, \forall w\, B[x, w]$$

Proof. Applied to the antecedent $\forall y^o w\, B[y, w] \to \perp$, Markov's principle $M_\wedge$ implies

(1) $\quad \exists XYW \neg\forall x < X\, B[Yx, Wx]$ which logically implies

(2) $\quad \exists XY \neg\forall x < X\forall w\, B[Yx, w]$.

Given X, Y, put $y := 1 + max\{Yx \mid x < X\}$. This y exemplifies, as a consequence of (2)

$\exists y \neg\forall x < y\, \forall w\, B[x, w]$ which is the succedent of the lemma.

We now come to

2.3.6 *Proof of theorem 2.3.3*

For any formula A of $L(HA^\omega)$, we show

(1) $\quad \forall x^o X^{o \to \tau} \neg\neg \exists Y^\tau\, A[x, X|x, Y]$ implies

(2) $\quad \neg\neg\exists F^{o \to \tau}\forall x^o\, A[x, F|x, Fx]$

Let

(3) $A^\wedge[x, X, Y] \equiv \exists v\forall w A_\wedge[x, X, Y, v, w]$.

Then (1) is in $HA^{\omega+}$ equivalent to

(4) $\quad \forall xX \neg\neg\exists Y v\forall w\, A_\wedge[x, X|x, Y, v, w]$

If, in proposition 2.3.4, we substitute $\forall w\, A_\wedge[x, X, Y, v, w]$ for $A[x, X, Y]$, hence also the tuple Y, v for the tuple Y, then (4), by that proposition, implies

(5) $\quad \neg\forall FV \exists y \neg\forall x < y\forall w A_\wedge[x, F|x, Fx, Vx, w]$

This implies by lemma 2.3.5

(6) $\quad \neg\forall FV \neg\forall yw A_\wedge[y, F|y, Fy, Vy, w]$ or equivalently

(7) $\quad \neg\neg\exists FV\forall yw A_\wedge[y, F|y, Fy, Vy, w]$ which implies

(8) $\quad \neg\neg\exists F\forall y\exists v\forall w A_\wedge[y, F|y, Fy, v, w]$

hence, in view of (3), also (2).

So, CDC_τ^- is proved without recourse to any stability postulates. This completes the proof of theorem 2.3.3 and, as remarked above, also the proofs of theorem 2.3.2 and of the $\wedge$-interpretation theorem 2.3.1.

Since by proposition 1.9.10

$$PA_0^{\omega*} \equiv PA_0^\omega + \{A \leftrightarrow A^{D-}\} \equiv PA_0^\omega + \{A \leftrightarrow A^{\wedge-}\} \equiv PA_0^{\omega+}$$

and since the argument leading to the $\wedge$-interpretation theorem 2.3.1 above remains correct if restricted to the type o fragments of the theories in question, we obtain as corollaries of theorems 2.3.2 and 2.3.1:

2.3.7   *Theorem (Howard [23])*

$$PA_0^{\omega *} + CDC_\tau^- \prec HA_0^{\omega +} + BI_\tau$$

2.3.8   *Theorem: Dialectica-interpretation of classical analysis (Spector [38], Howard [23])*

$$PA_0^{\omega *} + CDC_\tau^- \overset{D}{\hookrightarrow} T_0 + SimBR_\tau$$

Howard and Kreisel [25] also revert theorem 2.3.7 by proving BI_τ^- in $PA_0^\omega + (T - EXT) + (AC + DC)^-$. They thus show that, on the basis of $PA_0^\omega + (T - EXT)$, bar-induction and the classical choice principle $(AC + DC)^-$ are equivalent. Diller and Vogel [18] extend this result to the corresponding theories with language in all finite types by showing:

2.3.9   *Theorem*

$$PA^\omega + CDC_\sigma^- \vdash BI_\sigma^-$$

The *proof* consists in the construction of an unsecured infinite path under the assumption that the conclusion of BI_σ^- does not hold. We use the notation of 2.2.1 and abbreviate - for the sake of legibility - $\neg \forall z \neg$ by $\exists z$. In PA^ω, $Hyp\,4$ implies

$$\forall x X^{o \to \sigma} \exists Y^\sigma (\neg Q[X|x, x] \to \neg Q[X *_x Y, suc\,x])$$

Applying CDC_σ^- to the formula $\neg Q[X, x] \to \neg Q[X *_x Y, suc\,x]$, this implies by lemma 2.1.2

$$\exists F \forall x (\neg Q[F|x, x] \to \neg Q[F|suc\,x, suc\,x])$$

This yields by an application of (IND)

$$\neg Q[0^{o \to \sigma}, 0] \to \exists F \forall x \neg Q[F|x, x]$$

so that

$$Hyp\,3 \to \neg\, Q[0^{o \to \sigma}, 0] \to \neg\, Hyp\,1$$

which by contraposition yields $CDC_\sigma^- \to BI_\sigma^-$.

2.4 Strong computability of bar recursion

After it has been shown in the preceding sections that bar recursive functionals are an adequate tool for functional interpretations of intuitionistic as well as of classical analysis, we will now analyze the constructive content of these functionals, thus extending the work on normalization and computability of primitive recursive functionals in section 1.5. As discussed by Tait [44], since bar recursion recurs over well-founded trees, infinite sequences of functionals of a given type σ, i.e. *paths* in the tree $\mathbb{B}_\sigma$ (cf. background 2.1.4), must be considered in the computing process, e.g. for constructing a model of bar recursion. For that reason, we also restrict the presentation to functionals. Concerning the choice of the new functionals and their conversions, we follow Vogel [50].

2.4.1 *Bar recursors, infinite paths, and their conversions*

In extension of definition 2.1.5, **bar recursive functionals** are the applicative closure of the functionals of T and

(1) constants: bar recursors $\mathsf{B}_{\sigma\tau}$ and $\mathsf{B}^*_{\sigma\tau}$ for all finite types σ, τ

(2) infinite paths $\langle...\rangle$: If $a_0, ..., a_i, ...$ is an infinite sequence of bar-recursive functionals of type σ, then the path $\langle a_0, ..., a_i, ...\rangle$ is a bar-recursive functional of type $o \to \sigma$.

The conversion relation *conv*, as defined in 1.5.1, is extended by clauses for bar recursors

(B) $\qquad \mathsf{B}_{\sigma\tau}YGHctu \qquad conv \quad \mathsf{B}^*_{\sigma\tau}YGHct(Y(c *_t u) \dot{-} t)u$

(B^*0) $\qquad \mathsf{B}^*_{\sigma\tau}YGHct0u \qquad conv \quad G(c *_t u)t^+$

(B^*suc) $\qquad \mathsf{B}^*_{\sigma\tau}YGHcts^+u \qquad conv \quad H(\mathsf{B}_{\sigma\tau}YGH(c *_t u)t^+)(c *_t u)t^+$

and for infinite paths $\langle a_0, ..., a_i, ...\rangle : o \to \sigma$ and numerals n

$(\langle\rangle)$ $\qquad (\langle a_0, ..., a_i, ...\rangle n) \qquad conv \quad a_n$

The conversion rules for bar recursion stated here refer to types only and not to type tuples. That is no restriction, because the reduction process studied here includes product types which allow the derivation of $SimBR$ from bar recursion by an argument analogous to lemma 1.1.18 (cf. remark 2.1.10). Moreover, with some care, the following arguments may be read with type tuples instead of simple types.

The notions of contraction $\to_1$ and reduction $\to$, reduction tree, normalization, and strict and strong normalization are taken over from definitions

1.5.2 and 1.5.3 without change, of course restricted to functionals and based on the extended notion of conversion introduced here.

It may be observed that the reduction tree of a bar recursive functional is still a finitely branching tree, because there are no contractions within infinite sequences.

The conversions for bar recursors and infinite paths follow the pattern of conversions for primitive recursive terms introduced in 1.5.1 in that they do not involve any relations between the arguments of the converted term, in contrast to the conversions for the equality functionals introduced in 1.5.30. Therefore, concepts and results worked out in section 1.5 from definition 1.5.2 of contraction to definition 1.5.13 of strong computability $SComp_\tau$ and proposition 1.5.16 that strongly computable terms strongly normalize hold literally also for bar recursive functionals.

Also the argument for the continuity theorem 1.5.28 transfers to bar recursive functionals:

2.4.2 *Corollary to theorem 1.5.28: Continuity of strictly normalizing bar recursive functionals*

Strictly normalizing bar recursive functionals are continuous:
Let $Y : (o \to \sigma) \to o$ and $f : o \to \sigma$ be bar recursive functionals such that Yf stricly normalizes (to a numeral n). Then

$$\exists x^o \forall g^{o \to \sigma}(f|x = g|x \to Yf = Yg)$$

The *proof* of the continuity theorem 1.5.28 transfers completely to the present situation, as only one strict reduction chain from Yf to n is considered, and cases (i) and (ii) extend without change to all bar recursive functionals.

As the special case $g = f|y$ of this continuity result, we state:

2.4.3 *Corollary*

For Y, f as above,

$$\exists x^o \forall y^o (y > x \to Yf = Y(f|y))$$

Strong computability extends immediately to infinite sequences:

2.4.4 *Lemma*

If the bar recursive functionals $u_0, ..., u_i, ...$ are strongly computable of type σ, then also the infinite path $\langle u_0, ..., u_i, ... \rangle : o \to \sigma$ is strongly computable.

Proof. Any strongly computable functional $t : o$ strongly normalizes to a numeral $n = t^N$. So for any strongly computable functional $t : o$, there is a numeral n such that

$$\langle u_0, ..., u_i, ...\rangle t \to \langle u_0, ..., u_i, ...\rangle n \; conv \; u_n \in SComp_\sigma,$$

and that is the only way to convert such a functional. Therefore, together with all u_n, also the infinite sequence $\langle u_0, ..., u_i, ...\rangle$ is strongly computable.

The method of proving strong computability in theorem 1.5.17, however, also provides an important step towards strong computability of bar recursion:

2.4.5 *Lemma*

Let Y, G, H, c, t, u be strongly computable functionals.
Then $\mathsf{B}_{\sigma\tau}YGHctu \in SComp_\tau$, if
(i) $\quad Y(c *_t u) \le t$, *hence* $(Y(c *_t u) \dot{-} t)^N = 0$, *or*
(ii) $\quad \mathsf{B}_{\sigma\tau}YGH(c *_t u)(suct) \in SComp_{\sigma\to\tau}$

Proof. For some type tuple ρ, $\tau = \rho \to o$ or $\tau = \rho \to \tau_0 \times \tau_1$. Let $d \in SComp_\rho$ and put $r :\equiv \mathsf{B}YGHctud$. Any reduction chain of r contains a contraction

$$\mathsf{B}Y'G'H'c't'u'd' \to_1 \mathsf{B}^*Y'G'H'c't'(Y'(c' *_{t'} u') \dot{-} t')u'd'$$

where
(*) $\quad Y, G, H, c, t, u, d$ reduce to $Y', G', H', c', t', u', d'$
so that $(Y(c *_t u) \dot{-} t)^N = (Y'(c' *_{t'} u') \dot{-} t')^N$.
If (i) holds, then in any reduction chain of r the subterm $Y(c *_t u) \dot{-} t$ reduces to 0 ; so it will also contain a contraction

$$\mathsf{B}^*Y'G'H'c't'0u'd' \to_1 G'(c' *_{t'} u')t'^+d'$$

where, as above, (*) holds.
If (ii) holds and (i) does not, then in any reduction chain of r the subterm $Y(c *_t u) \dot{-} t$ reduces to a functional $suc\,s$ so that it will contain a contraction

$$\mathsf{B}^*Y'G'H'c't's^+ud' \to_1 H'(\mathsf{B}Y'G'H'(c' *_{t'} u')t'^+)(c *_{t'} u')t'^+d'$$

where again (*) holds.
So in both cases, r strongly normalizes, and if r reduces to a pair (e, f), a by (i) or (ii) strongly computable functional will do so too, and r as well as $\mathsf{B}YGHctu$ is strongly computable.

2.4.6   Proposition

Let Y, G, H, c_0, t_0 be strongly computable functionals and assume that $\mathsf{B}_{\sigma\tau}YGHc_0t_0$ is not strongly computable. Then there exists a strongly computable path $c : o \to \sigma$ in the tree $\mathbb{B}_\sigma$ running through the node $c_0|t_0, t_0$, i.e. $c|t_0 = c_0|t_0$, such that for all $n > t_0$

$\neg^\forall(i) \qquad Y(c|n) > n$

$\neg^\forall(ii) \qquad \mathsf{B}YGH(c|n)n \notin SComp_{\sigma\to\tau}$

Proof. For any $t \geq t_0$ and any finite sequence $c|t$ with $c|t_0 = c_0|t_0$, if $\mathsf{B}YGHcn \notin SComp_{\sigma\to\tau}$ for all n such that $t_0 \leq n \leq t$, then, by lemma 2.4.5, there is a $u \in SComp_\sigma$ such that the negations of (i) and (ii) in lemma 2.4.5 hold:

$\neg(i) \qquad Y(c *_t u) > n$

$\neg(ii) \qquad \mathsf{B}YGH(c *_t u)(suc\,t) \notin SComp_{\sigma\to\tau}$

By the axiom of dependent choice CDC_σ, there is an infinite path $c : o \to \sigma$ with $c|t_0 = c_0|t_0$ such that for all $n \geq t_0 \ \neg^\forall(i)$ and $\neg^\forall(ii)$ hold. This path c is strongly computable by lemma 2.4.4, because it is built up from strongly computable functionals c_0, t_0 and infinitely many $u \in SConp_\sigma$.

The conclusion of this proposition is clearly in conflict with corollary 2.4.3 on continuity:

2.4.7   Theorem: Strong computability of bar recursive functionals (Vogel [50])

All bar recursive functionals are strongly computable. They therefore also
- *strongly normalize*
- *strictly normalize*
- *are continuous.*

Proof. Since the functionals Y, c are strongly computable, Yc strictly normalizes to a numeral, and by corollary 2.4.3, there is a numeral n_0 such that

$$Y(c|n) = Yc \text{ for all } n > n_0$$

So $\neg^\forall(i)$ does not hold for $n > max(Yc)^N n_0$. Therefore, if Y, G, H, c_0, t_0 are strongly computable functionals, $\mathsf{B}_{\sigma\tau}YGHc_0t_0$ must be strongly computable, too. So $\mathsf{B}_{\sigma\tau}$ and then, of course, also $\mathsf{B}^*_{\sigma\tau}$ is strongly computable. The remaining claims of the theorem now follow by propositiion 1.5.16 and corollary 2.4.2 to theorem 1.5.28.

Chapter 3

Set Theory

Functional interpretations of set theoretic systems were started only in the 1990's, on a suggestion by A. Weiermann to attempt a Dialectica interpretation of Kripke–Platek set theory. We consider functional interpretations of classical Kripke–Platek as well as of Constructive Zermelo–Fraenkel set theory, both in finite types.

In section 3.1, we present an intuitionistic theory $T_\in$ of constructive set functionals, essentially following Burr [6], together with its classical extension $T_\in^c$. Whereas in arithmetic theories, we avoid any application of extensionality principles, it seems necessary to include weak extensionality in the form of the rule $(T - EXT)$ already in the theory $T_\in$.

In section 3.2, Kripke–Platek set theory in finite types $KP\omega^\omega$ is then introduced as an extension of $T_\in^c$, and Dialectica- as well as $\wedge$-translation transfer in a natural way from the $\exists$free language of classical arithmetic PA^ω to the $\exists$free language of $KP\omega^\omega$. However, whereas the full theory $KP\omega^\omega$ is $\wedge$-interpreted in $T_\in^c$, a Dialectica-interpretation even of its type o fragment $KP\omega_0^\omega$ is possible only in an extension of $T_\in^c$ by a classical, non-constructive uniform choice functional.

The $\wedge$-interpretation of Constructive Zermelo–Fraenkel set theory $CZF^{\omega-}$ in section 3.3 may be considered the central result of this chapter. It requires a translation of the existential quantifier – which does not occur in languages of classical theories treated in the previous section – different from its translation in the arithmetic case. An adequate translation was initiated by Burr [6] and completed by Schulte [35]. A simplified version (cf. [13]) is given under definition 3.3.3 as an addition to the $\wedge$-translation of the language of $KP\omega^\omega$. The $\wedge$-interpretation of $CZF^{\omega-}$, as far as the negative fragment, Δ_0-separation and, in particular, transfinite induction are concerned, can be taken over from the $\wedge$-interpretation of $KP\omega^\omega$ in section 3.2; however, there remains considerable work in $\wedge$-interpreting laws for disjunction and existence and, in particular, the schema of *strong collection*. Due to the "weak" translation of the existential quantifier, two of the three schemata characterizing $\{A \leftrightarrow A^\wedge\}$ must be put into a "weak" form (cf. 3.3.15).

Modified realization mr of $CZF^{\omega-}$ is developed, together with its t- and q-hybrids, in section 3.4. As in the arithmetic case, all three realizations have the same characterization, but in contrast to the arithmetic case, they are characterized by schemata of weak choice and of weak independence of premise. Also the q- and t-hybrids of the $\wedge$-interpretation of $CZF^{\omega-}$ are developed in parallel in section 3.5. They also have the same characterization; however, as in the arithmetic case, this characterization differs widely from the characterization of the $\wedge$-interpretation of $CZF^{\omega-}$ mentioned above. In the final section 3.6, it is shown that the axiom of extensionality $(t-ext)$ is not $\wedge$-interpretable in $T_\in$: All functionals of $T_\in$ are majorized by constructive set functionals, though not all of them by functionals in $T_\in$, whereas functionals $\wedge$-interpreting $(t-ext)$ are not majorizable by any functionals.

3.1 Constructive set functionals

Constructive set functionals, in particular primitive recursive set functions, have been studied, among others, by Jensen-Karp [26], Rathjen [33], and Aczel-Rathjen [2]. An intuitionistic theory $T_\in$ of constructive set functionals was developed by Burr [6]. Like the theories T and $T_\wedge$, $T_\in$ is an extension of the basic logic of typed terms $BLTT$ in section 0.1, however, in linear types only.

3.1.1 *Description of the theory $T_\in$ of constructive set functionals and its classical and type o versions*

The theory $T_\in$ is an extension of $BLTT$, formulated in linear types.

(1) Basic symbols of $T_\in$, beyond those of $BLTT$, are the primitive 2-place relation symbol $\in$, the disjunction symbol $\vee$, bounded quantifiers $\forall x \in t$ and $\exists x \in t$, and constants that will be given below together with their defining axioms. Terms are built up as in $BLTT$, including, of course, the new constants, and excluding pairing and projection functionals.

(2) Within the language of $T_\in$, we distinguish the class of Δ_0-*formulae*:

1. Expressions $s \in t$, equations $s = t$ between terms $s, t : o$, and $\perp$ are (atomic) Δ_0-formulae.

2. If A, B are Δ_0-formulae, then $A \to B$, $A \wedge B$, and $A \vee B$ are Δ_0-formulae.

3. If A is a Δ_0-formula and the term $t : o$ does not contain the variable $x : o$, then $\forall x \in t\, A$ and $\exists x \in t\, A$ are Δ_0-formulae.

(3) This class of Δ_0-formulae is extended to the full language $L(T_\in)$ as follows:

1. Δ_0-formulae and equations $a = b$ between terms $a, b : \tau \neq o$ are $T_\in$-formulae.

2. If A, B are $T_\in$-formulae, then $A \to B$ and $A \wedge B$ are $T_\in$-formulae.

3. If A is a $T_\in$-formula and the term $t : o$ does not contain the variable $x : o$, then $\forall x \in t\, A$ is a $T_\in$-formula.

There are only *bounded* formulae in $L(T_\in)$, and disjunctions and bounded existential quantifiers occur only in Δ_0-formulae.

We write $t\#0$ and read "t is *inhabited*" for $\exists x \in t\; x = x$, and we occasionally abbreviate $t\#0 \wedge \forall x \in t$ by $\forall x \in t\#0$.

(4) Axioms and rules of $BLTT$ are extended by axioms $A8$ and rule $A9$ for disjunction (cf. 1.2.2) and the following axioms and rules for bounded quantification:

$Q1b$. $\forall x \in t\, A[x] \to s \in t \to A[s]$ for terms $s, t : o$ with s substitutible in $A[x]$;

$Q2b$. $s \in t \wedge A[s] \to \exists x \in t\, A[x]$ for terms s, t as above;

$Q3b$. $A \to x \in t \to B \vdash A \to \forall x \in t\, B$, if $x : o$ is not free in A;

$Q4b$. $x \in t \wedge A \to B \vdash \exists x \in t\, A \to B$, if $x : o$ is not free in B.

Hence, the logic of $T_\in$ is intuitionistic predicate logic with bounded quantification only.

(5) New constants – beyond those of $BLTT$ – are:

1. arithmetic constants (besides $0 : o$) $\mathsf{Suc} : o \to o \to o$, $\omega : o$ with axioms

$(0)\ \neg\, x \in 0$

That is, 0 denotes the empty set $\emptyset$.

$(\mathsf{Suc})\ x \in \mathsf{Suc}\, st \leftrightarrow x \in s \vee x = t$

which shorthand may be written $\mathsf{Suc}\, st = s \cup \{t\}$,

i.e. we have a **two-place successor** operation, and the **one-place successor** $suc\, t$ is defined as $\mathsf{Suc}\, tt$.

$(\omega)\ x \in \omega \leftrightarrow (x = 0 \vee \exists y \in \omega\; x = suc\, y)$

So, as a consequence of (T IND) (cf. (6) below), ω denotes the set of *numerals*, and we write $1 = suc\,0$, $2 = suc\,1$ etc.

2. constants for **union-replacement, intersection, and implication**

$\mathsf{U} : (o \to o) \to o \to o$, $\mathsf{Int} : o \to o \to o \to o$, $\mathsf{Imp} : o \to o \to o \to o$, with axioms

(U) $x \in \mathsf{U}ft \leftrightarrow \exists y \in t\; x \in fy$, shorthand
$$\mathsf{U}ft = \bigcup \{fy \mid y \in t\}$$

(Int) $x \in \mathsf{Int}\,ts \leftrightarrow x \in s \wedge \forall y \in t\; x \in y$, shorthand
$$\mathsf{Int}\,ts = s \cap \bigcap t$$

(Imp) $x \in \mathsf{Imp}\,rst \leftrightarrow x \in r \wedge (x \in s \to x \in t)$, shorthand
$$\mathsf{Imp}\,rst = \{x \in r \mid x \in s \to x \in t\}$$

3. a **transfinite** or $\in$-**recursor** $\mathsf{R}_\tau : ((o \to \tau) \to o \to \tau) \to o \to \tau$ for any linear type τ with axiom

(TR) $\mathsf{R}_\tau ht = h((\mathsf{R}_\tau h) \restriction_\tau t)t$

The *restriction functional* $\restriction_\tau$ will be defined below in 3.1.4, independent, of course, of the recursor R_τ.

(6) Further axioms and rules of $T_\in$ are the Δ_0-axiom of **set-extensionality**

(ext) $(\forall x \in s\; x \in t \wedge \forall x \in t\; x \in s) \to s = t$,

the rule $(T - EXT)$ of **type-extensionality** as formulated under 0.1.14, for all formulae A of $L(T_\in)$, and finally the rule of **transfinite** or $\in$-**induction**

(T IND) $\forall x \in y\; F[x] \to F[y] \vdash F[t]$ for all $T_\in$-formulae $F[y]$.

It should be clear that the axiom (ext) is a set-theoretic axiom, whereas the rule $(T - EXT)$ is a type-theoretic, hence logical rule which does not bear any information on the $\in$-relation.

This completes the description of the theory $T_\in$.

(7) The *classical version* $T_\in^c$ of $T_\in$ is $T_\in + Stab(=)$, i.e.
$$T_\in^c :\equiv T_\in + \{\neg\neg a = b \to a = b : a, b : \tau, \tau\ a\ linear\ type\}$$

(8) The *type o versions* $T_{\in 0}$ of $T_\in$ and $T_{\in 0}^c$ of $T_\in^c$ are extensions of $BLTT_0$ (in linear types) and the maximal fragments of $T_\in$ and $T_\in^c$, respectively, in Δ_0-language as described above under item (2), with axioms (TR) "read under context" (like (K) and (S)) and type extensionality in the form of the rule $(T - EXT_0)$ as formulated for linear types in 0.1.14.

$T_\in$ is an intuitionistic theory of constructive set functionals that has bounded quantifiers only and may therefore be called *bounded* or *almost quantifier-free*. As the bounds may be infinite, $T_\in$ is by no means a finitary theory. If only ω is accepted as a bound and if R_τ is accepted only for $\tau = o$ and only restricted to ω, unbounded Heyting arithmetic HA is yet contained in this minute fragment of $T_\in$, even of $T_{\in 0}$.

We define some standard set-theoretic operations in this theory.

3.1.2 *Definition of elementary set operations*

241

(1) $\{t\} := \mathsf{Suc}\, 0t$

(2) $\{s, t\} := \mathsf{Suc}\{s\}t = \mathsf{Suc}(\mathsf{Suc}\, 0s)t$

(3) $s \cap t := \mathsf{Imp}\, sst$

(4) $s - t := \mathsf{Imp}\, st0$

(5) $\bigcup t := \mathsf{U}(\lambda x.x)t$

(6) $s \cup t := \bigcup\{s, t\} = \mathsf{U}(\lambda x.x)\{s, t\}$

(7) $\{fx \mid x \in t\} := \mathsf{U}(\lambda x.\{fx\})t$

(8) $\{fxy \mid y \in Yx, x \in X\} := \mathsf{U}(\lambda x.\{fxy \mid y \in Yx\})X$

(9) $\{x \in r \mid s[x] \in t[x]\} := \{\mathsf{K}xy \mid y \in \{s[x]\} \cap t[x], x \in r\}$

(10) $s \cap \bigcap\{fx \mid x \in t\} := \mathsf{Int}\,\{fx \mid x \in t\}s$

Also beyond this list, we use further standard set theoretic notation, e.g.

(11) $s \subset t :\Leftrightarrow \forall x \in s\ x \in t$

3.1.3 *Lemma: Elementary properties of set operations*

In $T_\in$ is provable:

(1) $x \in \{t\} \leftrightarrow x = t$

(2) $x \in \{s, t\} \leftrightarrow x = s \vee x = t$

(3) $x \in s \cap t \leftrightarrow x \in s \wedge x \in t$

(4) $x \in s - t \leftrightarrow x \in s \wedge \neg\, x \in t$

(5) $x \in \bigcup t \leftrightarrow \exists y \in t\ x \in y$

(6) $x \in s \cup t \leftrightarrow x \in s \vee x \in t$

(7) $y \in \{fx \mid x \in t\} \leftrightarrow \exists x \in t\ y = fx$

(8) $z \in \{fxy \mid y \in Yx, x \in X\} \leftrightarrow \exists x \in X\ \exists y \in Yx\ z = fxy$

(9) $z \in \{x \in r \mid s[x] \in t[x]\} \leftrightarrow z \in r \wedge s[z] \in t[z]$

(10) $y \in s \cap \bigcap\{fx \mid x \in t\} \leftrightarrow y \in s \wedge \forall x \in t\ y \in fx$

Proof. The calculations are straight forward. We show (9):

$z \in \{x \in r \mid s[x] \in t[x]\} \leftrightarrow z \in \{\mathsf{K}xy \mid y \in \{s[x]\} \cap t[x], x \in r\} \leftrightarrow$
$\exists x \in r\, \exists y \in \{s[x]\} \cap t[x]\,(z = x) \leftrightarrow z \in r \wedge \exists y \in t[z]\, y = s[z] \leftrightarrow$
$z \in r \wedge s[z] \in t[z]$ by (8), (K), (3), and (1).

Already on the basis of union-replacement (U), (1) and (3), a *restriction functional* $\restriction_\tau$ can be defined.

3.1.4 *Recursive definition of restriction functionals $\restriction_\tau$ of type $(o \to \tau) \to o \to o \to \tau$ for linear types τ*

We write $(f \restriction_\tau t)$ or just $(f \restriction t)$ for $\restriction_\tau ft$ and put

$(f \restriction_o t)x = \bigcup f(\{x\} \cap t)$
$(f \restriction_{\sigma \to \tau} t)xu^\sigma = ((\lambda y.fyu^\sigma) \restriction_\tau t)x$

This definition fills the gap in the definition (TR) of transfinite recursion.

3.1.5 *Lemma: Laws of restriction*

The restriction functional $\restriction_\tau$ satisfies provably in $T_\in$:

(i) $\forall x \in t\,(f \restriction t)x = fx$
(ii) $\forall x \in t\, fx = gx \to f \restriction t = g \restriction t$, *moreover*
(iii) $s \subset t \to (f \restriction t) \restriction s = f \restriction s$

Proof of (i) and (ii) by induction on τ.

1. $(\tau = o)$ $\forall x \in t\,(f \restriction_o t)x = \bigcup\{fz \mid z \in \{x\} \cap t\} = \bigcup\{fx\} = fx$,

hence (i); also

$$\forall x \in t\, fx = gx \to (f \restriction t)y = \bigcup\{fx \mid x \in \{y\} \cap t\}$$
$$= \bigcup\{gx \mid x \in \{y\} \cap t\} = (g \restriction t)y,$$

and by an application of rule $(T - EXT)$, (ii) follows.

2. (i): $\forall x \in t\,(f \restriction_{\sigma \to \tau} t)xu^\sigma = ((\lambda y.fyu^\sigma) \restriction_\tau t)x$
which by I.H. $= (\lambda y.fyu^\sigma)x = fxu^\sigma$
(ii): $\forall x \in t\, fx = gx \to (f \restriction_{\sigma \to \tau} t)xu^\sigma = ((\lambda y.fyu^\sigma) \restriction_\tau t)x$
which by I.H. $= ((\lambda y.gyu^\sigma) \restriction_\tau t)x = (g \restriction_{\sigma \to \tau} t)xu^\sigma$
In either case, the result follows by $(T - EXT)$.

Proof of (iii). Let $s \subset t$. By (i),

$$\forall x \in s\,((f \restriction t) \restriction s)x = (f \restriction t)x = fx = (f \restriction s)x,$$

and $(f \restriction t) \restriction s = f \restriction s$ follows by (ii).

It is this kind of construction that makes us accept type-extensionality $(T-EXT)$ in $T_\in$. In Gödel's T, an arithmetic restriction functional is defined in 1.4.1 by primitive recursion (of higher type) combinatorially, without using $(T - EXT)$. We conjecture that an analogous, purely combinatorial construction in $T_\in$ is impossible. Another simple application of $(T - EXT)$ is the case distinction functional for higher types:

3.1.6 Lemma: Case distinction

For any linear type τ, there is a case distinction functional
$D_\tau : \tau \to \tau \to o \to \tau$ *satisfying for arbitrary $a, b : \tau$*

$$T_\in \vdash D_\tau ab0 = a \,\wedge\, (x\#0 \to D_\tau abx = b)$$

Proof by induction on τ. We define:
1. $D_o abx := \bigcup(\{a \mid x \in 1\} \cup \{b \mid z \in x\})$
2. $D_{\rho \to \sigma} abx := \lambda z^\rho . D_\sigma (az^\rho)(bz^\rho)x$
Then, the induction hypothesis and $(T - EXT)$ prove the lemma.

On the basis of the set forming constants U, Int, Imp, the set forming operations defined in 3.1.2 quickly lead to:

3.1.7 Proposition: Explicit Δ_0-separation

Given a Δ_0-formula $F[x]$ and a term $t : 0$ not containing x, there exists a term $\{x \in t \mid F[x]\}$ of $L(T_\in)$ such that

$$T_\in \vdash y \in \{x \in t \mid F[x]\} \leftrightarrow y \in t \wedge F[y]$$

The *proof* is given by simply stating a recursive definition of separation terms along the inductive definition of Δ_0-formulae. The correctness of our definition follows by lemma 3.1.3 and several distributive laws that hold in intuitionistic logic.
$\{x \in t \mid \bot\} = 0$
$\{x \in t \mid r \in s\} = \{\mathsf{K}xy \mid y \in \{r\} \cap s, x \in t\}$ is (9) in Lemma 3.1.3.
$\{x \in t \mid r = s\} = \{x \in t \mid r \in \{s\}\}$
$\{x \in t \mid F[x] \wedge G[x]\} = \{x \in t \mid F[x]\} \cap \{x \in t \mid G[x]\}$
$\{x \in t \mid F[x] \vee G[x]\} = \{x \in t \mid F[x]\} \cup \{x \in t \mid G[x]\}$
$\{x \in t \mid F[x] \to G[x]\} = \mathsf{Imp}\, t\{x \in t \mid F[x]\}\{x \in t \mid G[x]\}$
For the bounded quantifier cases $\exists/\forall y \in s[x]\, F[x, y]$, there is, by I.H., a Δ_0-term $\{z \in t \mid F[z, y]\}$, and by applications of (9) in 3.1.3, we get
$\{x \in t \mid \exists y \in s[x]\, F[x, y]\} = \{x \in t \mid x \in \bigcup\{\{z \in t \mid F[z, y]\} \mid y \in s[x]\}\}$
$\{x \in t \mid \forall y \in s[x]\, F[x, y]\} = \{x \in t \mid x \in t \cap \bigcap\{\{z \in t \mid F[z, y]\} \mid y \in s[x]\}\}$

By induction on Δ_0-formulae $F[x]$, the proposition follows.

It is primarily for this result that Burr [6] introduces the functionals Int and Imp as primitives into the theory $T_\in$. It immediately yields:

3.1.8   Corollary: Existence of characteristic terms

Every Δ_0-formula A has a characteristic term $\{0|A\}$ such that

$$T_\in \vdash A \leftrightarrow \{0|A\} = 1$$

This term is, of course, $\{x \in 1 \mid A\}$ with x not free in A.

This corollary underlines the analogy between the functional theories $T_\in$ and $T_\wedge$ of section 1.1: Up to provable equivalence, the language of $T_\in$ is built up from equations of arbitrary type and $\bot$ solely by conjunction, implication, and bounded universal quantification, like the language of $T_\wedge$. However, the proof of this corollary makes use of more complex Δ_0-axioms in which disjunctions and bounded existential quantifiers occur. So it seems natural that, for $T_\in$, full Δ_0-language and -logic is used, in contrast to the situation in $T_\wedge$. An important difference in the logic of the two systems remains: Whereas in the arithmetical theory $T_\wedge$, equations of type o are decidable, in set theoretic $T_\in$ this is not the case, despite the existence of characteristic terms. The characteristic terms, however, suffice to show that the above definition of the classical version $T_\in^c$ of $T_\in$ is adequate:

3.1.9   Definition and corollary: Stability and restricted excluded middle in $T_\in^c$

(1) *$T_\in^c$ proves stability for the full $T_\in$-language:*

Stab $\neg\neg A \to A$

(2) *Moreover, $T_\in^c$ proves the schema of restricted excluded middle:*

(rem) $A \vee \neg A$ *for Δ_0-formulae A*

(3) *On the basis of $T_{\in 0}$, all the schemata Stab(=), Stab, (rem), and the general law of excluded middle are equivalent.*

Proof. By the previous corollary, every Δ_0-formula is in $T_\in$ equivalent to an equation of type o. Therefore, every $T_\in$-formula is in $T_\in$ equivalent to a negative formula built up from equations and $\bot$. For such formulae, stability follows from $Stab(=)$.

That proves (1). And since $\neg\neg(A \vee \neg A)$ is provable already in theories with constructive logic like $T_\in$, (2) follows. (3) is the obvious restriction of (1) and (2) to Δ_0-language.

It may be stated without proof that in $T_\in$, (rem) does not imply $Stab(=)$; thus $T_\in + (rem)$ is a weaker theory than $T_\in^c$.

We now define a (coding of) ordered pairs different from the Kuratowski pair $\{\{x\}, \{x, y\}\}$, because the corresponding decoding operations for this pair are more uniformly definable.

3.1.10 Definition of singleton, ordered pair, tuple coding and decoding, disjoint sum

Consider the Δ_0-property of a set u being a *singleton*,

$$Sing(u) :\equiv u \# 0 \land \forall x \in u \, \forall y \in u \; x = y$$

An *ordered pair* $\langle \;,\; \rangle$ and *decoding operations* $(\;)_0$ and $(\;)_1$ are then defined by

$$\langle x, y \rangle := \{\{x\}, \{0, \{y\}\}\}$$

$(z)_0 := \bigcup\bigcup\{u \in z \mid Sing(u)\}$ and $(z)_1 := \bigcup\bigcup\bigcup\{u \in z \mid \neg Sing(u)\}$.

These pairing and decoding functions are extended to n-place codings $\langle \;,...,\; \rangle^n$ and decoding functions $(\;)_i^n$ for $i < n$, $n > 0$ by
$\langle x \rangle^1 = x$ and $\langle x_0, ..., x_n \rangle^{n+1} = \langle \langle x_0, ..., x_{n-1} \rangle^n, x_n \rangle$ for $n > 0$,
$(z)_n^{n+1} = (z)_1$ and $(z)_i^{n+1} = (z)_i^n$ for $i < n$.

The *disjoint sum* of the family of sets $\{Yx \mid x \in X\}$ is the set

$$\{\langle x, y \rangle \mid y \in Yx, x \in X\}$$

3.1.11 Lemma: Decoding of ordered pairs and tuples

(1) $T_\in \vdash (\langle x, y \rangle)_0 = x \land (\langle x, y \rangle)_1 = y$
(2) $T_\in \vdash \langle x, y \rangle^2 = \langle x, y \rangle$
(3) $T_\in \vdash (\langle x_0, ..., x_n \rangle^{n+1})_i = x_i$ for $i \leq n$

Proof. First, for $x, y : o$, $\langle x, y \rangle$ is a set by (1) and (2) of 3.1.2, and for $z : o$, $(z)_0$ and $(z)_1$ are sets by explicit Δ_0-separation 3.1.7 and (5) of 3.1.2. The disjoint sum is the special case $f = \langle \;,\; \rangle$ of 3.1.2 (8).
(1) Since $T_\in$ proves $Sing(\{x\})$ and $\neg Sing(\{0, \{y\}\})$, $T_\in$ also proves
$\{u \in \langle x, y \rangle \mid Sing(u)\} = \{\{x\}\}$ and $\{u \in \langle x, y \rangle \mid \neg Sing(u)\} = \{\{0, \{y\}\}\}$.
Therefore

$$(\langle x, y \rangle)_0 = \bigcup\bigcup\{\{x\}\} = \bigcup\{x\} = x \text{ and}$$

$(\langle x, y \rangle)_1 = \bigcup \bigcup \bigcup \{\{0, \{y\}\}\} = \bigcup \bigcup \{0, \{y\}\} = \bigcup (0 \cup \{y\}) = y$

(2) $\langle x, y \rangle^2 = \langle \langle x \rangle^1, y \rangle = \langle x, y \rangle$

(3) follows by induction on n.

On the logical side, this provides a contraction rule for bounded universal quantifiers:

3.1.12   Lemma: Contraction of bounds

In $T_{\in}$ is provable

$$\forall x \in X \forall y \in Yx \; F[x, y] \leftrightarrow \forall z \in \{\langle x, y \rangle \mid y \in Yx, x \in X\} \, F[(z)_0, (z)_1]$$

$$X \# 0 \wedge \exists x \in X \; Yx \# 0 \; \leftrightarrow \; \{\langle x, y \rangle \mid y \in Yx, x \in X\} \# 0$$

Proof of the first statement. For the implication from left to right, let $z \in \{\langle x, y \rangle \mid y \in Yx, x \in X\}$. Then $(z)_0 \in X$ and $(z)_1 \in Y(z)_0$. So, the lefthand side implies $F[(z)_0, (z)_1]$. Conversely, let $x_0 \in X$ and $y_0 \in Yx_0$. Then $\langle x_0, y_0 \rangle \in \{\langle x, y \rangle \mid y \in Yx, x \in X\}$. Thus, the righthand side implies $F[x_0, y_0]$ by the above lemma.
The second statement is immediate.

We now discuss several principles of recursion and induction used in the sequel. The functionals of coding and decoding are, in the presence of type extensionality $(T - EXT)$, the tool to derive simultaneous transfinite recursion from the schema (TR).

3.1.13   Proposition: Simultaneous transfinite recursion

Given a type tuple $\tau = \tau_0, ..., \tau_n$ and terms $g_i : (o \to \tau) \to o \to \tau_i$ $(i \leq n)$, there are terms $f_i : o \to \tau_i$ $(i \leq n)$ which solve the $n + 1$ equations

$$(SimTR) \qquad f_i t = g_i (f_0 \upharpoonright t)...(f_n \upharpoonright t)t \qquad (i \leq n)$$

Proof. For $n = 0$, $\tau = \tau_0$, and the above equation is solved by $f_0 = \mathsf{R}_\tau g_0$. Let $n > 0$ and $i \leq n$. The linear types τ_i are of the form $\sigma_i \to o$ for some type tuples σ_i. Let u_i be a tuple of fresh variables of type tuple σ_i so that $f_i t u_i$ are terms of type o. Moreover, put

$$\sigma \equiv \sigma_0, ..., \sigma_n, \quad u \equiv u_0, ..., u_n, \quad \text{and} \quad u^i \equiv 0^{\sigma_0}, ..., u_i, ..., 0^{\sigma_n}$$

Thus, the tuple u^i of tuples consists of the tuple u_i, flanked by n tuples of 0-functionals. Now let

$$ftu = \langle f_0 t u_0, ..., f_n t u_n \rangle$$

Then, for $i \leq n$, if $y \in t$, we have
$$f_i y u_i = (fyu)_i = (fyu^i)_i = ((f \restriction t)yu^i)_i$$
which implies
$$(f_i \restriction t)xu_i = \bigcup\{f_i y u_i \mid y \in \{x\} \cap t\} = \bigcup\{(f \restriction t)yu^i \mid y \in \{x\} \cap t\}$$
$$= h_i(f \restriction t)txu_i$$
for $h_i = \lambda z t x u_i . \bigcup\{zyu^i \mid y \in \{x\} \cap t\}$. By $(T - EXT)$,
$$f_i \restriction t = h_i(f \restriction t)t$$
The f_i that solve $(SimTR)$ must therefore satisfy
$$f_i t = g_i(h_0(f \restriction t)t)...(h_n(f \restriction t)t)t = G_i(f \restriction t)t$$
for $G_i = \lambda z t . g_i(h_0 z t)...(h_n z t)t$ $(i \leq n)$. However, the schema (TR) provides a term f which solves
$$ft = \lambda u . \langle G_0(f \restriction t)tu_0, ..., G_n(f \restriction t)tu_n \rangle$$
Since $f_i t u_i = (ftu)_i$, we obtain
$$f_i t = G_i(f \restriction t)t = g_i(f_0 \restriction t)...(f_n \restriction t)t$$
as desired.

It may be noticed that $fxu = \langle f_0 x u_0, ..., f_n x u_n \rangle$ implies
$$(f \restriction t)xu = \langle (f \restriction t)xu_0, ..., (f \restriction t)xu_n \rangle$$
only for $x \in t$, whereas for $x \notin t$ and $n > 0$ the last equation does not hold. This problem is avoided in the presence of product types. The argument presented here may, however, be extended to show conservativity of $T_\in$ with product types over $T_\in$ in linear types only.

In section 1.4, in the proof of 1.4.4, we gave a combinatorial derivation of simultaneous course-of-values recursion $(SimCV)$ from course-of-values recursion (CV). The argument given there apparently transfers to the present set theoretic context only for *transitive* restriction terms t. Together with ω-induction, the concept of transitivity is used here to derive the principle of ω-recursion.

3.1.14 *Proposition: ω-induction*

In $T_\in$ is admissible
$$F[0], \forall y \in \omega \ (F[y] \rightarrow F[suc\,y]) \vdash t \in \omega \rightarrow F[t]$$
Proof. By axiom (ω), if $x \in \omega$, either $x = 0$, and by the first hypothesis, we have $x = 0 \rightarrow F[x]$, or $\exists y \in \omega \ x = suc\,y$ which, combined with the second hypothesis, may be written $\exists y \in x \ (y \in \omega \land x = suc\,y \land (F[y] \rightarrow F[x]))$. So, in either case,
$$\forall y \in x \ (y \in \omega \rightarrow F[y]) \rightarrow (x \in \omega \rightarrow F[x])$$
By (T IND), the assertion follows.

3.1.15   *Definition of transitivity*

A set t is *transitive*, *tran(t)*, if $\forall x \in t \, \forall y \in x \; x \in t$, i.e.

$$tran(t) \Leftrightarrow \forall x \in t \; x \subset t \Leftrightarrow \bigcup t \subset t$$

3.1.16   *Corollary: Transitivity of numerals and of ω*

In $T_\in$ is provable

    (i) $tran(t) \rightarrow tran(suc\,t) \wedge \bigcup(suc\,t) = t$
    (ii) $\forall x \in \omega \; tran(x)$
    (iii) $tran(\omega)$

Proof. (i) If $x \in suc\,t$, either $x \in t$ or $x = t$, so, due to $tran(t)$, in either case, $x \subset t \subset (suc\,t)$. So, $tran(suc\,t)$ as well as $\bigcup(suc\,t) = t$.
(ii) $tran(0)$ holds vacuously. By (i) and ω-induction 3.1.14, (ii) follows.
(iii) $0 \subset \omega$ and $\forall x \in \omega(x \subset \omega \rightarrow suc\,x = x \cup \{x\} \subset \omega)$.
Hence, by ω-induction, $tran(\omega)$.

3.1.17   *Proposition: Simultaneous ω-recursion*

Given a non-empty type tuple $\tau = \tau_0, ..., \tau_n$ and terms $a_i : \tau_i$ and $b_i : \tau \rightarrow o \rightarrow \tau_i$ $(i \leq n)$, there are terms $f_i : o \rightarrow \tau_i$ $(i \leq n)$ which solve the $n + 1$ pairs of equations

$$f_i 0 = a_i \quad and \quad \forall x \in \omega \; f_i(suc\,x) = b_i(f_0 x)...(f_n x)x$$

Proof. We use the case distinction functionals D_{τ_i} of 3.1.6 and define $f_0, ..., f_n$ by simultaneous transfinite recursion 3.1.13:

$$f_i y = D_{\tau_i} a_i (b_i((f_0 \upharpoonright y)\textstyle\bigcup y)...((f_n \upharpoonright y)\bigcup y)\bigcup y)$$

Then $f_i 0 = a_i$ and for $y = suc\,x \in \omega$, $y \# 0$ and $\bigcup y = x \in y$ by the last corollary, so that by lemma 3.1.5 (i) for $x \in \omega$

$$f_i(suc\,x) = b_i((f_0 \upharpoonright (suc\,x))x)...((f_n \upharpoonright (suc\,x))x)x = b_i(fx)x$$

A principle of generalized induction holds in $T_\in$ which is the essential technical tool for a functional interpretation of (T IND). As an alternative to Burr [6], we give a proof of this principle closely related to the proof of the corresponding principle 1.1.22 in $T_\wedge$.

3.1.18   Proposition: Generalized transfinite induction

Given a term X, a term tuple W with variables $a, u, x : o$ and a tuple of variables z, all not in X, W, such that

(1) $T_\in \vdash \forall u \in a \, \forall x \in Xaz \, B[u, Wxaz] \to B[a, z]$

then $T_\in \vdash B[t, z]$ for any term $t : o$.

Proof. Let a term tuple X_1, Z be defined by simultaneous ω-recursion 3.1.17:

$$Zazy0 = z \qquad \forall n \in \omega \, Zaz\langle x, y \rangle (suc\,n) = Wxa(Zazyn)$$

$$X_1az0 = 1 \qquad \forall n \in \omega \, X_1az(suc\,n) = \{\langle x, y \rangle \mid x \in Xa(Zazyn), y \in X_1azn\}$$

By contraction of bounds 3.1.12, these terms satisfy

(2) $\forall n \in \omega \forall u \in a(\forall y \in X_1az(suc\,n)B[u, Zazy(suc\,n)] \leftrightarrow$
$\qquad\qquad \forall y \in X_1azn \forall x \in Xa(Zazyn) \, B[u, Wxa(Zazyn)]])$

(1), with $Zazyn$ substituted for z and $\forall y \in X_1azn$ distributed over the implication, may be rewritten under this equivalence as

$\forall n \in \omega(\forall u \in a \forall y \in X_1az(suc\,n) \, B[u, Zazy(suc\,n)] \to \forall y \in X_1azn \, B[a, Zazyn])$

Distributing $\forall n \in \omega$ over this implication and strengthening the antecedent slightly by including in it the case $n = 0$, this implies

$\forall u \in a \forall n \in \omega \forall y \in X_1azn \, B[u, Zazyn] \to \forall n \in \omega \forall y \in X_1azn \, B[a, Zazyn]$

From this, we conclude by (T IND)

$$\forall n \in \omega \forall y \in X_1tzn \, B[t, Ztzyn])$$

which for $n = 0$ implies $T_\in \vdash B[t, z]$, as claimed.

3.2   Kripke–Platek set theory and its functional interpretations

Kripke–Platek set theory with infinity $KP\omega$ (cf. Barwise [4]) is a classical set theory which lacks strong non-constructive principles like the powerset axiom or full separation, but still has a considerable expressive power. In a sense, $KP\omega$ collects constructive set theoretic principles in a classical theory. We underline this aspect by $\wedge$-interpreting $KP\omega$, rather its finite type extension $KP\omega^\omega$, in the classical theory $T_\in^c$ of constructive set functionals introduced in the last section.

3.2.1   Description of Kripke–Platek set theory in finite (linear) types $KP\omega^\omega$ as an extension of $T_\in^c$

(1) Basic symbols of $KP\omega^\omega$ are those of $T_\in$, plus the universal quantifier $\forall u^\tau$ for all linear types τ.

(2) $KP\omega^\omega$ has the same Δ_0-language as $T_\in$. The full language of $KP\omega^\omega$ is the closure of this Δ_0-language and all equations $a = b$ with $a, b : \tau$ under conjunction $\wedge$, implication $\rightarrow$, bounded and unbounded universal quantification $\forall x \in t$ and $\forall u^\tau$. In other words, the formulae of $KP\omega^\omega$ are inductively defined like those of $T_\in$ in 3.1.1(3), with the additional clause

4. If A is a formula of $KP\omega^\omega$ and u^τ is a variable of type τ, then $\forall u^\tau A$ is a formula of $KP\omega^\omega$.

(3) Axioms and rules of $KP\omega^\omega$ are those of $T_\in^c$, extended – with the exception of the rule $(T - EXT)$ – to the full language of $KP\omega^\omega$, plus the following:

1. the laws Q1 and Q3 for the universal quantifier (cf. 1.2.1);
2. the axiom schema of Δ_0-collection:

$$(\Delta_0\text{-}collection) \quad \forall x \in t \, \neg\forall y \neg A[x,y] \rightarrow \neg\forall z \neg \forall x \in t \, \exists y \in z \, A[x,y]$$
$$\text{where } A[x,y] \text{ is a } \Delta_0\text{-formula and } x, y : o$$

The rule $(T - EXT)$ of type-extensionality remains restricted to the language of $T_\in$ (*weak extensionality*):

$(T - EXT) \quad A \rightarrow fx = gx \vdash A \rightarrow f = g$
$\qquad$ for A in $L(T_\in)$, x not in A, f, g

The rule of transfinite induction (T IND), however, is extended to the full language:

(T IND) $\quad \forall x \in y \, F[x] \rightarrow F[y] \vdash F[t] \quad$ for all $KP\omega^\omega$-formulae $F[y]$

Being an extension of $T_\in^c$, $KP\omega^\omega$ takes over several properties from this subtheory:

3.2.2   Lemma

(1) $KP\omega^\omega$ enjoys explicit Δ_0-separation.
(2) In $KP\omega^\omega$, every formula is equivalent to a negative formula.
(3) $KP\omega^\omega \vdash \neg\neg A \rightarrow A$ for all formulae A of $KP\omega^\omega$.
(4) $KP\omega^\omega$ satisfies classical predicate logic.

This follows from proposition 3.1.7 and by the same argument by which corollary 3.1.9 is shown.

3.2.3 Lemma

$KP\omega^\omega$ *is an extension of* $KP\omega$:

$$KP\omega \vdash A \quad \text{implies} \quad KP\omega^\omega \vdash A$$

Proof. It suffices to prove the axioms of $KP\omega$ (see Barwise [4]) in $KP\omega^\omega$. The axiom of Extensionality is, as (ext), already an axiom of $T_\in$. In the negative fragment, the axiom schema of Foundation takes the form

$$\forall x(\forall y \in x\, F[y] \to F[x]) \to F[t]$$

which follows by the rule (T IND), if this rule is applied to this longer formula.

Explicit versions of the axioms of Pair, Union and the schema of Δ_0-Separation are proved already in $T_\in$ (cf. 3.1.3 (2) and (5), and proposition 3.1.7).

The schema of Δ_0-Collection is an axiom schema of $KP\omega^\omega$, in fact the only axiom schema extending $T_\in$.

Under the proviso that functional translations of $KP\omega^\omega$ translate the language of $T_\in$ identically, it is clear how the Dialectica- as well as the $\wedge$-translation are transferred from the arithmetic case to $KP\omega^\omega$: For D, the recursive clauses are literally the same as in the arithmetic case; for $\wedge$, the symbol $<$ in $(\to)^\wedge$ is replaced by $\in$. To avoid ambiguities, we give a simultaneous definition of D and $\wedge$ on $L(KP\omega^\omega)$.

3.2.4 Recursive definition of the Dialectica- and the $\wedge$-translation on $L(KP\omega^\omega)$

Let I stand for the translations D and $\wedge$ simultaneously. To any formula A of $KP\omega^\omega$, I assigns an expression $A^I \equiv \exists v \forall w A_I[v, w]$, with $A_I[v, w]$ a formula of $L(T_\in)$.

$(T_\in)^I \quad A^I \equiv A \quad \text{for} \quad A \in L(T_\in)$

Let A^I be as above and $B^I \equiv \exists y \forall z B_I[y, z]$, then

$(\wedge)^I \quad (A \wedge B)^I \quad \equiv \exists vy \forall wz(A_I \wedge B_I)$

$(\to)^D \quad (A \to B)^D \quad \equiv \exists WY \forall vz(A_D[v, Wvz] \to B_D[Yv, z])$

$(\to)^\wedge \quad (A \to B)^\wedge \quad \equiv \exists XWY \forall vz(\forall x \in Xvz\, A_\wedge[v, Wxvz] \to B_\wedge[Yv, z])$

  in case the tuple w is not empty

$(\to)_0^\wedge \quad (A \to B)^\wedge \quad \equiv \exists Y \forall vz(A_\wedge[v] \to B_\wedge[Yv, z])$ for empty w

$(\forall \in)^I \quad (\forall x \in t\, B[x])^I \equiv \exists Y \forall z(\forall x \in t\, B_I[x, Yx, z])$

$(\forall)^I \quad (\forall u A[u])^I \quad \equiv \exists V \forall uw A_I[u, Vu, w]$

It is to be expected that, as in the arithmetic case, the Dialectica interpretation does not interpret $A7$ for formulae of higher type. However, in interpreting set theory, the Dialectica interpretation runs into a problem already with type o formulae:

3.2.5 The $\Delta_0 \subset \Pi_1$-problem

Given a $T_\in$-formula $A[y^o]$, the formula $\forall y \in t\, A[y]$ - which is Δ_0 in case $A[y]$ is Δ_0 - may be written as a Π_1-formula $\forall y(y \in t \to A[y])$, and
$$KP\omega^\omega \vdash \forall y(y \in t \to A[y]) \to \forall y \in t\, A[y]$$
The D-translation of this formula calls for a term s satisfying
$$(s \in t \to A[s]) \to \forall y \in t\, A[y]$$
Such a term may not exist in $T_\in$, as the example $A[y] \equiv \bot$ shows:

Example. $KP\omega$ (with a constant 0 for the empty set) proves
$$\forall x(\forall y \lnot y \in x \to x = 0)$$
This has the D-translation
$$\exists Y \forall x(\lnot Yx \in x \to x = 0)$$
Any Y satisfying this formula is necessarily a classical choice functional which is not a constructive functional and hence not in $T_\in^c$.

The $\wedge$-translation of the above formula, however, is (equivalent to)
$$\exists Y(\forall y \in Y(y \in t \to A[y]) \to \forall y \in t\, A[y])$$
and is therefore $\wedge$-interpreted by $Y = t$:

The $\wedge$-translation solves the $\Delta_0 \subset \Pi_1$-problem, whereas the D-translation does not solve it, at least not in $T_\in^c$.

Here we made use of a simplification of the $\wedge$-translation:

3.2.6 Lemma

A formula $\forall y\, A[y] \to B$ with A, B in $L(T_\in)$, $y : o$ may be $\wedge$-translated as
$$\exists Y(\forall y \in Y\, A[y] \to B)$$
without changing the concept of $\wedge$-interpretability.

Proof. $(\forall y\, A[y] \to B)^\wedge$ is literally $\exists XY'(\forall x \in X\, A[Y'x] \to B)$. Y is obtained from X, Y' by putting $Y = \{Y'x \mid x \in X\}$, and X, Y' are obtained from Y by $X = Y$ and $Y'x = x$. Moreover, in the translation of longer formulae, the tuple of variables X, Y' and the variable Y are handled the same way.

3.2.7 ∧-*Interpretation theorem for* $KP\omega^\omega$

Kripke–Platek set theory in finite types is ∧-*interpretable in* $T_\in^c$:

$$KP\omega^\omega \overset{\wedge}{\hookrightarrow} T_\in^c$$

Proof by induction on deductions. For the axioms and rules of logic concerning the negative particles, the proof parallels the proof in the arithmetic case in 1.2.7. We take the ∧-translations over from there, replacing $<$ by $\in$, use the same induction hypotheses as there, and check the properties of $T_\in$ used to construct the interpreting terms.

A1, more generally the schema $A \to B$ with $A^\wedge \equiv B^\wedge$, A4, A5, A⊥, the equality rule (Eq), the universal quantifier laws $Q1$ (including $Q1b$) and $Q3$ are interpreted exactly as in the arithmetic case, using, besides logic, essentially only the fact $\forall x \in 1\, A[x] \to A[0]$. $Stab(=)$ is interpreted trivially. A2 is ∧-interpretable, because $\vdash \forall w A[w]$ implies $\vdash \forall x \in t\, A[Wx]$ for any W and t.

A3. Given interpreting term tuples X_1, W_1, Y_1 and X_2, W_2, Y_2 for the hypotheses $A \to B$ and $B \to C$, and abbreviating $t[x_2] := W_2 x_2 (Y_1 v) z$ and $s := X_2 (Y_1 v) z$ as in 1.2.7, we obtain by contraction of bounds 3.1.12

$$\forall x \in Xvz\, A_\wedge[v, W_1(x)_0 vt[(x)_1]] \to C_\wedge[Y_2(Y_1 v), z]$$

where $Xvz := \{\langle x_1, x_2 \rangle \mid x_1 \in X_1 vt[x_2], x_2 \in s\}$
So, this term X and tuples W, Y, given by

$$Wxvz = W_1(x)_0 v(W_2(x)_1(Y_1 v)z) \text{ and } Yv = Y_2(Y_1)v,$$

∧-interpret $A \to C$.

A6. Given term tuples X_1, W_1, X_2, W_2, Y interpreting $A \to (B \to C)$, the tuples X with $Xvz = X_1 vz \cup X_2 vz$, $W = W_1, W_2$, and Y interpret $(A \wedge B) \to C$.

A7. Given term tuples X_1, W_1, Y_1 and X_2, W_2, Y_2 ∧-interpreting $A \to B$ and $A \to C$ respectively, tuples X, W, Y_1, Y_2 with

$$Xvz_1 z_2 = \{\langle x_0, 0 \rangle \mid x_0 \in X_1 vz_1\} \cup \{\langle x_0, 1 \rangle \mid x_0 \in X_2 vz_2\},$$

the disjoint union of $X_1 vz_1$ and $X_2 vz_2$, and

$$\begin{aligned} Wxvz_1 z_2 &= W_1(x)_0 vz_1 \quad if \quad (x)_1 = 0, \\ &= W_2(x)_0 vz_2 \quad if \quad (x)_1 \,\#\, 0 \end{aligned}$$

∧-interpret $A \to B \wedge C$.
A8 and $Q2b$ are Δ_0-formulae and are therefore ∧-interpreted trivially.

$Q3b.$ $A \to y \in t \to B[y] \vdash A \to \forall y \in t\, B[y]$

After abstracting from y, we are given X_0, W_0, Y such that

$$\forall x \in X_0 yvz\, A_\wedge[v, W_0 xyvz] \to y \in t \to B_\wedge[y, Yyv, z]$$

Distributing $\forall y \in t$ over this implication, we obtain by contraction of bounds 3.1.12 for $Xvz := \{\langle x, y \rangle \mid x \in X_0 yvz, y \in t\}$

$$\forall x \in Xvz A_\wedge[v, W_0(x)_0(x)_1 vz] \to \forall y \in t\, B_\wedge[y, Yyv, z]$$

so that X as just defined, $Wxvz := W_0(x)_0(x)_1 vz$, and Y as given, $\wedge$-interpret the conclusion.

We come to the rules introducing disjunction and bounded existential quantification in the antecedent.

$A9, Q4b$. If these inferences are in $T_\in$, the interpretation is trivial. If not, then the consequent formula C is not in $L(T_\in)$, and we use the following tautologies, provable in $T_\in^c$, to avoid these inferences:

$$A \vee B \leftrightarrow \neg(\neg A \wedge \neg B) \text{ and } \exists x \in t\, A[x] \leftrightarrow \neg \forall x \in t \neg A[x]$$

From $A \to C$ and $B \to C$ with A, B in Δ_0, we derive $\neg(\neg A \wedge \neg B) \to C$ in the negative fragment, possibly using $Stab(=)$, and compute a $\wedge$-interpreting term tuple Y for this formula. Y then also $\wedge$-interprets $A \vee B \to C$. Similarly, from $x \in t \wedge A[x] \to B$ with A in Δ_0 and x not free in B, we derive $\neg \forall x \in t \neg A[x] \to B$ in the negative fragment and compute a $\wedge$-interpreting term tuple Y for this formula. Y then also $\wedge$-interprets $\exists x \in t\, A \to B$.

Since the non-logical axioms and rules of $T_\in^c$, including $(T - EXT)$, are $\wedge$-interpreted trivially, it remains to interpret Δ_0-collection and (T IND).

The $\wedge$-translation of Δ_0-collection (cf. 3.2.1 (3)) is, applying lemma 3.2.6,

$$\exists Z \forall Y (\forall x \in t\, \exists y \in Yx\, A[x, y] \to \exists z \in ZY \forall x \in t\, \exists y \in z\, A[x, y])$$

Given Y satisfying the antecedent, there is one canonical z satisfying the consequent, i.e. $z = \bigcup Yt$. (Δ_0-collection) is therefore $\wedge$-interpreted by the term Z with the value $ZY = \{\bigcup Yt\}$.

(T IND) $\forall u \in a\, F[u] \to F[a] \vdash F[t]$

By I.H., there are terms and term tuples X, W, Y_0 such that

(1) $T_\in^c \vdash \forall x \in Xavz\, \forall u \in a\, F_\wedge[u, vu, Wxavz] \to F_\wedge[a, Y_0 va, z]$

We define terms Y by simultaneous transfinite recursion 3.1.13

$$Ya = Y_0(Y \upharpoonright a)a,$$

substitute $Y \upharpoonright a$ for v in (1), and obtain

$$T^c_{\in} \vdash \forall x \in Xa(Y \upharpoonright a)z \, \forall u \in a \, F_\wedge[u, Yu, Wxa(Y \upharpoonright a)z] \to F_\wedge[a, Ya, z]$$

Here, the terms $Xa(Y \upharpoonright a)z$, $Wxa(Y \upharpoonright a)z$ are terms $X'az, W'xaz$, and $F_\wedge[a, Ya, z]$ is a formula $B[a, z]$ satisfying (1) in proposition 3.1.18. So, by generalized transfinite induction, $F_\wedge[t, Yt, z]$ follows.
This completes the proof of the $\wedge$-interpretation theorem. For related proofs using, however, different translations, cf. Burr [6], [7], Schulte [35].

An immediate consequence is:

3.2.8   Corollary: Conservativity and relative consistency

$KP\omega^\omega$ is a conservative extension of $T^c_{\in}$. The consistency of $T^c_{\in}$ implies the consistency of $KP\omega^\omega$.

As stated above, this proof does not transfer to a Dialectica interpretation in $T^c_{\in}$, even of $KP\omega$ only. We therefore extend the theory $T^c_{\in}$ accordingly.

3.2.9   The axiom of uniform choice

Let F_C be a new constant of type $o \to o$ with axiom

(uniform AC) $x \# 0 \to \mathsf{F}_C x \in x$

$Th + $ (uniform AC) then stands for the extension of a theory Th by the constant F_C and the axiom (uniform AC).

The extension of $T^c_{\in 0}$ by (uniform AC) suffices for a Dialectica interpretation of $KP\omega$, even of the type o fragment $KP\omega^\omega_0$ of $KP\omega^\omega$:

3.2.10   Dialectica interpretation theorem

$KP\omega^\omega_0 + $ (uniform AC) is D-interpretable in $T^c_{\in 0} + $ (uniform AC):

$$KP\omega^\omega_0 + (uniform\ AC) \overset{D}{\hookrightarrow} T^c_{\in 0} + (uniform\ AC)$$

The proof by induction on deductions parallels the proof of the corresponding interpretation theorem 3.2.7 for the $\wedge$-interpretation and is, in some aspects, technically simpler. In particular, the induction steps for $A3$ and (T IND) are special cases of the corresponding steps in 3.2.7, with the given bounds $X_1v_1z_1, X_2v_2z_2$ resp. $Xavz$ being equal to 1. Two induction steps, however, those for contraction $A7$ and for introduction of the bounded universal quantifier $Q3b$, require new arguments.

A7. By I.H., there are term tuples W_1, Y_1 and W_2, Y_2 D-interpreting $A \to B$ and $A \to C$, i.e.

$$\vdash A_D[v, W_1 v z_1] \to B_D[Y_1 v, z_1] \text{ and } \vdash A_D[v, W_2 v z_2] \to C_D[Y_2 v, z_2]$$

We look for a term tuple W with

$$\vdash (B_D[Y_1 v, z_1] \to W v z_1 z_2 = W_2 v z_2) \wedge (\neg B_D[Y_1 v, z_1] \to W v z_1 z_2 = W_1 v z_1),$$

because, for such a W,

$\vdash \neg B_D[Y_1 v, z_1] \to A_D[v, W v z_1 z_2] \to \bot$ and
$\vdash B_D[Y_1 v, z_1] \to A_D[v, W v z_1 z_2] \to C_D[Y_2 v, z_2]$.
Hence, by classical logic,

$$\vdash A_D[v, W v z_1 z_2] \to B_D[Y_1 v, z_1] \wedge C_D[Y_2 v, z_2]$$

Since B_D is Δ_0, B_D has the characteristic term $\{0 \mid B_D[Y_1 v, z_1]\}$, by corollary 3.1.8. Therefore, by lemma 3.1.6, the term tuple W defined by

$$W v z_1 z_2 = D_\tau (W_1 v z_1)(W_2 v z_2)\{0 \mid B_D[Y_1 v, z_1]\}$$

with D_τ a tuple of case distinction functionals of appropriate type tuple τ, and Y_1, Y_2 will D-interpret $A \to B \wedge C$. - This argument, cut back to Heyting arithmetic HA, is in fact Gödel's original argument for his Dialectica interpretation of A7.

$Q3b$. By I.H., there are term tuples W', Y such that
$\vdash A_D[v, W'vyz] \to y \in t \to B_D[y, Yvy, z]$.
By Frege's rule 0.1.6, this implies
$\vdash \forall y \in t\, A_D[v, W'vyz] \to \forall y \in t\, B_D[y, Yvy, z]$.
If the antecedent of this implication does not hold, then we pick a $y \in t$ such that $\neg A_D[v, W'vyz]$ by applying the choice functional F_C, that is, we define

$$W vz = W'v(\mathsf{F}_C\{y \in t \mid \neg A_D[v, W'vyz]\})z$$

and obtain

$$\vdash \neg \forall y \in t\, A_D[v, W'vyz] \to A_D[v, Wvz] \to \bot$$

By classical logic, these two implications imply

$$\vdash A_D[v, Wvz] \to \forall y \in t\, B_D[y, Yvy, z]$$

(uniform AC) is D-interpreted trivially. That completes the Dialectica interpretation of $KP\omega_0^\omega + $ *(uniform AC)*.

It is now routine to transfer arguments and results from classical arithmetic PA^ω in section 1.9 to the present situation in $KP\omega^\omega$.

3.2.11   Axioms of choice for $KP\omega^\omega$

The *quantifier-free axiom of choice* is, in analogy to 1.9.5,

$$(qf - AC) \qquad \forall x \neg \forall y \neg A[x,y] \to \neg \forall Y \neg \forall x\, A[x, Yx]$$

with $A[x,y]$ in $L(T_\in)$. This is, as in 1.9.5, essentially of the form

$$B \to B^{D-} \text{ with } B \equiv (\forall x \exists y\, A[x,y])^-$$

The *quantifier free axiom of restricting choice* now reads

$$(qf - ARC) \qquad \forall x \neg \forall y \neg A[x,y] \to \neg \forall S, Y \neg \forall x \neg \forall s \in Sx \neg A[x, Ysx]$$

with $A[x,y]$ also from $L(T_\in)$. This is, again as in 1.9.5, of the form

$$B \to B^{\wedge -} \text{ with } B \equiv (\forall x \exists y\, A[x,y])^-$$

The statements of lemma 1.9.6, with PA^ω replaced by $KP\omega^\omega$, now follow by the same proofs, for $\wedge$ as well as for D:

3.2.12   Lemma

Let I stand for $\wedge$ or for D. For $A, B, A[u] \in L(KP\omega^\omega)(+F_C)$, u a - possibly empty - tuple of variables, $KP\omega^\omega$ proves:

1. $A^{I-} \leftrightarrow A$ for $A \in L(T_\in)$
2. $(A \wedge B)^{I-} \leftrightarrow A^{I-} \wedge B^{I-}$
3. $(A \to B)^{I-} \to A^{I-} \to B^{I-}$
4. $(\forall u \neg A[u])^{I-} \to \forall u \neg (A[u]^{I-})$; *furthermore*
5. $KP\omega^\omega + (qf - ARC) \vdash \forall u \neg (A[u]^{\wedge -}) \to (\forall u \neg A[u])^{\wedge -}$
6. $KP\omega^\omega + (qf - AC) \vdash \forall u \neg (A[u]^{D-}) \to (\forall u \neg A[u])^{D-}$

For *prenex formulae* $B \equiv \forall v_1 \neg ... \forall v_n \neg C[v]$ with $C[v] \in L(T_\in)$, this implies by subformula induction:

3.2.13   Lemma: Partial characterizations of translations $\wedge$ and D in $KP\omega^\omega$

On the basis of $KP\omega^\omega$, the schemata

$$(qf - ARC) \text{ and } \{B \leftrightarrow B^{\wedge -} \mid B \text{ prenex}\}$$

as well as the schemata

$$(qf - AC) \text{ and } \{B \leftrightarrow B^{D-} \mid B \text{ prenex}\}$$

are equivalent:

$$KP\omega^\omega + (qf - ARC) \equiv KP\omega^\omega + \{B \leftrightarrow B^{\wedge -} \mid B \text{ prenex}\}$$
$$KP\omega^\omega + (qf - AC) \equiv KP\omega^\omega + \{B \leftrightarrow B^{D-} \mid B \text{ prenex}\}$$

Together with the interpretation theorems 3.2.7 and 3.2.10, these partial characterizations yield:

3.2.14　Characterization theorem for the ∧-translation on $KP\omega^\omega$

(1) *On the basis of* $KP\omega^\omega$, $(qf - ARC)$ *and* $\{A \leftrightarrow A^{\wedge-}\}$ *are equivalent:*

$$KP\omega^\omega + (qf - ARC) \equiv KP\omega^\omega + \{A \leftrightarrow A^{\wedge-}\}$$

(2) *Any formula* ∧-*interpretable in* $T_\in^c$ *is derivable in* $KP\omega^\omega + (qf - ARC)$.

3.2.15　Characterization theorem for the D-translation on $KP\omega_0^\omega + (uniform\ AC)$

(1) *On the basis of* $KP\omega_0^\omega + (uniform\ AC)$, $(qf - AC)$ *and* $\{A \leftrightarrow A^{D-}\}$ *are equivalent:*

$$KP\omega_0^\omega + (uniform\ AC) + (qf - AC) \equiv KP\omega_0^\omega + (uniform\ AC) + \{A \leftrightarrow A^{D-}\}$$

(2) *Any formula* D-*interpretable in* $T_{\in 0}^c$ *is derivable in*

$$KP\omega_0^\omega + (uniform\ AC) + (qf - AC)$$

For a *proof* of both of these theorems, cf. the proof of theorem 1.9.7, the corresponding characterization of ∧ on PA^ω.

The interpretation theorems above extend maximally as in section 1.9, inverting the inferences (2) of the characterization theorems:

3.2.16　Extended ∧- and D-interpretation theorems

$$KP\omega^\omega + (qf - ARC) \overset{\wedge}{\hookrightarrow} T_\in^c$$

$$KP\omega_0^\omega + (uniform\ AC) + (qf - AC) \overset{D}{\hookrightarrow} T_{\in 0}^c + (uniform\ AC)$$

Proof. Extending theorems 3.2.7 and 3.2.10, it suffices to ∧-interpret $(qf - ARC)$ and to D-interpret $(qf - AC)$ in $T_\in^c$. Arithmetic versions of these interpretations are given in the proofs of theorems 1.9.9 and 1.9.11. In the ∧-interpretation of $(qf - ARC)$, of course, symbols $<$ must be replaced by symbols $\in$. The D-interpretation of $(qf - AC)$ in T is literally also a D-interpretation in $T_\in^c$.

Up to this point, there is a convincing analogy between interpretation and characterization theorems of classical set theory $KP\omega^\omega$ and classical arithmetic PA^ω, by replacing $<$ in arithmetic by $\in$ in set theory. That analogy holds in spite of the fact that $KP\omega^\omega$ is not just a negative version (plus stability) of some constructive set theory as PA^ω is of HA^ω. One may therefore ask: Why does the Dialectica interpretation of $KP\omega_0^\omega$ have a $\Delta_0 \subset \Pi_1$ problem which does not occur in arithmetic? The answer is simple: The analog of *(uniform AC)* in arithmetic is

$$x \neq 0 \rightarrow 0 < x \ :$$

the "uniform choice" in arithmetic is the number 0. Restricted to transitive sets, *(uniform AC)*, in analogy to the above, becomes the triviality

$$tran(x) \wedge x\#0 \rightarrow 0 \in x$$

So the point is that the arithmetical $<$-relation is transitive while the set theoretic $\in$-relation in general is not.

The analogy between classical arithmetic and classical set theory extends to dependence results concerning the choice schemata relevant in both cases, as stated in proposition 1.9.12.

3.2.17   *Proposition*

(1) $KP\omega^\omega + (qf - AC) \vdash (qf - ARC)$
(2) $KP\omega_0^\omega + (uniform\ AC) + (qf - ARC) \vdash (qf - AC)$

Proof. For (1), see 1.9.12 (1).
(2). As we work in $KP\omega_0^\omega$, the formula $A[x, y]$ in $(qf - AC)$ is Δ_0. Given $(\forall x \exists y A[x, y])^-$, by $(qf - ARC)$ we get

$$(\exists SY' \forall x \exists s \in Sx\, A[x, Y'sx])^-$$

so that, for these S, Y' and for all x, the Δ_0-separation term $\{s \in Sx \mid A[x, Y'sx]\}$ is inhabited. Applying *(uniform AC)*, we define

$$Yx = \mathsf{F}_C\{s \in Sx \mid A[x, Y'sx]\}$$

and obtain $(\exists Y \forall x A[x, Yx])^-$, as was to be shown.

It is plausible that further results from section 1.9 about PA_0^ω should also hold for $KP\omega_0^\omega$. However, we do not see how to transfer the proofs to the present situation. We therefore close the section with some conjectures.

3.2.18   *Conjectures*

(1) $KP\omega_0^\omega + (uniform\ AC) \nvdash (qf - AC)$
(2) $KP\omega_0^\omega + (uniform\ AC) \nvdash (qf - ARC)$
(3) $KP\omega_0^\omega + (uniform\ AC) \vdash A^{D-} \leftrightarrow A^{\wedge-}$

For the independence result (1), the argument for 1.9.12 (3) transfers to the present situation, if there is a minimal model of $KP\omega_0^\omega +$ *(uniform AC)*, call it MIN, as follows:
the constant ω is interpreted in MIN by the set of numerals;
the type 1 objects of MIN, i.e. the elements of $MIN_{0\to0}$, are interpretations in MIN of type 1 functionals of $T_\in^c + (uniform\ AC)$.
It would be interesting to know whether $KP\omega_0^\omega + (uniform\ AC)$ has such a model.
(2) follows from conjecture (1) by (2) of the last proposition.
(3) corresponds to proposition 1.9.10 for PA_0^ω which in turn follows from the corresponding statement for HA_0^ω. However, as mentioned above, there is no constructive set theory that corresponds to $KP\omega^\omega$ as HA^ω corresponds to PA^ω.

3.3   Constructive set theory in finite types $CZF^{\omega-}$ and its $\wedge$-interpretation

We change back to functional interpretations of constructive theories and consider a constructive theory of sets in linear types. This set theory, like Kripke–Platek set theory, avoids non-constructive principles. It does so, however, on the basis of intuitionistic logic, and on that basis, the principle of *strong collection* becomes a constructive principle. This may seem surprising at first sight, but it becomes convincingly clear by the $\wedge$-interpretation in the theory $T_\in$.

3.3.1   *Description of Constructive Zermelo–Fraenkel set theory in finite linear types $CZF^{\omega-}$ as an extension of $T_\in$*

(1) Basic symbols of $CZF^{\omega-}$ are those of $T_\in$, plus the unbounded quantifiers $\forall u^\tau, \exists u^\tau$ for all linear types τ.

(2) $CZF^{\omega-}$ has the same Δ_0-language as $T_\in$. The full language of $CZF^{\omega-}$ is the closure of this Δ_0-language and all equations $a = b$ with $a, b : \tau$ under conjunction $\wedge$, implication $\to$, bounded universal quantification $\forall x \in t$ and unbounded universal and existential quantification $\forall u^\tau, \exists u^\tau$. In other words, the formulae of $CZF^{\omega-}$ are inductively defined like those of $T_\in$ in 3.1.1 (3), with the additional clause

4. If A is a formula of $CZF^{\omega-}$ and u^τ is a variable of type τ, then

$\forall u^\tau A$ and $\exists u^\tau A$ are formulae of $CZF^{\omega-}$.

(3) Axioms and rules of $CZF^{\omega-}$ are those of $T_\in$, extended – with the exception of the rule $(T - EXT)$ – to the full language of $CZF^{\omega-}$, plus the following:

1. the quantifier laws Q1 to Q4 (cf. 1.2.1);
2. the axiom schema of strong collection:

(Strong collection) $\quad \forall x \in a \exists y F[x,y] \to \exists b F'[a,b] \quad\quad$ where

$$F'[a,b] \equiv \forall x \in a \exists y \in bF[x,y] \wedge \forall y \in b \exists x \in aF[x,y]$$

for all formulae $F[x,y]$ of $L(CZF^{\omega-})$.

The rule $(T - EXT)$ of type-extensionality (cf. 0.1.14) remains restricted to the language of $T_\in$ (*weak extensionality*), while the rule of transfinite induction (T IND) is extended to the full language:

(T IND) $\quad \forall x \in y\, F[x] \to F[y] \vdash F[t]$

for all formulae $F[y]$ of $L(CZF^{\omega-})$

As in $L(T_\in)$, disjunctions $\vee$ and bounded existential quantifiers $\exists x \in t$ occur only in Δ_0-formulae. If $F[x^o]$ is not a Δ_0-formula, $\exists x \in tF[x]$ will be read as an abbreviation for $\exists x(x \in t \wedge F[x])$. If A or B are not Δ_0-formulae, disjunction may be defined by

$$A \vee B :\equiv \exists x(x \in \omega \wedge (x = 0 \to A) \wedge (x \# 0 \to B))$$

3.3.2 Constructive Zermelo–Fraenkel set theory as a first order theory

A constructive set theory CZF was introduced by Myhill [31]. CZF^- denotes this theory in the formulation of Aczel [1] without the axiom schema of subset collection. As explicit versions of the axioms of Infinity (by (ω)), Pair (by (2)), Union (by (5) in Lemma 3.1.3) and of the schema of Δ_0-Separation (by proposition 3.1.7) are derivable already in $T_\in$, we see:

Lemma $\quad CZF^{\omega-}$ *is an extension of* CZF^-.

Except for the treatment of type-extensionality $(T - EXT)$ which makes $CZF^{\omega-}$ a weakly extensional theory, violating the deduction theorem, $CZF^{\omega-}$ is the *natural span* of CZF^- and $T_\in$.

We now turn to functional interpretations of $CZF^{\omega-}$. We take over the $\wedge$-translation from the previous section and amend it by a clause for the existential quantifier. This version of the $\wedge$-translation simplifies the $\times$-translations introduced by Burr [7] and Schulte [35], taking over Schulte's $\times$-translation of the existential quantifier.

3.3.3 *Recursive definition of the $\wedge$-translation on $L(CZF^{\omega-})$*

This is a literal repetition of definition 3.2.4 for the case $I = \wedge$, extended by the clause

$(\exists)^\wedge$ $\quad (\exists u A[u])^\wedge \;\equiv\; \exists XUV \forall w (X \# 0 \wedge \forall x \in X \; A_\wedge[Ux, Vx, w])$

There are two clauses by which existential quantifiers enter into the prefix of $A^\wedge$, viz. $(\to)^\wedge$ and $(\exists)^\wedge$. In both cases, these existential quantifiers are accompanied by a bound, Xvz in the first case and an inhabited X in the second case. In $(\to)^\wedge$, we call the variable X as well as the term Xvz an *accompanying bound* of w as well as of X, W. In $(\exists)^\wedge$, we call X an *accompanying bound* of u as well as of X, U, V. Then every existential variable in the prefix of $A^\wedge$ has at least one accompanying bound.

Due to $(\exists)^\wedge$, we do not always have $A^{\wedge\wedge} \equiv A^\wedge$, but one easily sees:

3.3.4 *Lemma*

In $CZF^{\omega-}$ is provable:

1. $(\forall w A)^\wedge$ $\quad \leftrightarrow \forall w A$ $\quad$ for $A \in L(T_\in)$
2. $(A \wedge B)^\wedge$ $\quad \leftrightarrow (A^\wedge \wedge B^\wedge)$
3. $(A \to B)^\wedge$ $\quad \to (A^\wedge \to B^\wedge)$
4. $(\forall x \in t \; A[x])^\wedge \to \forall x \in t (A[x]^\wedge)$
5. $(\forall u A[u])^\wedge$ $\quad \to \forall u (A[u]^\wedge)$
6. $(\exists u A[u])^\wedge$ $\quad \leftrightarrow \exists u (A[u]^\wedge)$
7. $A^{\wedge\wedge}$ $\quad \leftrightarrow A^\wedge$

Proof. 1. holds, because $(\forall w A)^\wedge \equiv \forall w A$ for $A \in L(T_\in)$. 2. to 5. are shown as in lemma 1.2.12 for the $\wedge$-translation of arithmetic. 6. is

$\exists XUV \forall w (X \# 0 \wedge \forall x \in X \; A_\wedge[Ux, Vx, w]) \leftrightarrow \exists u v \forall w \; A_\wedge[u, v, w]$

Given X, U, V satisfying the left hand side, take an $x \in X$ - which is possible, since X is inhabited for any w - , then Ux, Vx will be a tuple u, v satisfying the right hand side. Conversely, given u, v satisfying the right hand side, then the tuple $X = 1, U = \lambda x.u$, and $V = \lambda x.v$ will satisfy the left hand side.

7. follows from 1. by iterated application of 6.

3.3.5 The $\Delta_0 \subset \Sigma_1$-Problem

Before we tackle the $\wedge$-interpretation of $CZF^{\omega-}$, we give an argument for the above formulation of $(\exists)^{\wedge}$. Dual to the $\Delta_0 \subset \Pi_1$-problem discussed in the previous section, in the language of $CZF^{\omega-}$ we have a $\Delta_0 \subset \Sigma_1$-problem: a Δ_0-formula $\exists x \in t\, A[x]$ may also be read as a Σ_1-formula $\exists x (x \in t \wedge A[x])$ with an unbounded quantifier $\exists x$ in $L(CZF^{\omega-})$, having a different translation. In the simplest case $A[x] \equiv x = x$, i.e. $\exists x \in t\, A[x] \equiv t\#0$, a functional translation I as used in arithmetic would translate the implication $\quad B \equiv t\#0 \to \exists x\, x \in t \quad$ into

$$B^I \equiv \exists x (t\#0 \to x \in t)$$

An I-interpreting x would again - as with the $\Delta_0 \subset \Pi_1$-problem - lead to a uniform choice functional F_C with

$$t\#0 \to \mathsf{F}_C t \in t$$

which is not in $T_\in$. In this situation, the adequate translation is given in Burr [6] by his $\times$-translation: In translating $\exists x^o C[x]$ with $C[x] \in \Delta_0$, he basically looks for a set X such that

(1) X is inhabited;
(2) all elements $x \in X$ satisfy $C[x]$.

For the above B, this asks for a set X satisfying

$$B_\times \equiv B_\wedge \equiv t\#0 \to X\#0 \wedge \forall x \in X\, x \in t$$

Obviously, in this case, the set $X = t$ will do.
So it is just the introduction of an inhabited bound X and of the combination $X\#0 \wedge \forall x \in X$ into the $\times$-translation of the existential quantifier that solves the $\Delta_0 \subset \Sigma_1$-problem. This example also shows that functional interpretations are not about formal equivalences, but about finding explicit instances (in parameters) that satisfy a given existential formula. In our $\Delta_0 \subset \Sigma_1$-example, we do find $X = t$ satisfying $B_\times$, but we do not find a term x satisfying B_I, in spite of the formal equivalence of $\exists X B_\times$ and $\exists x B_I$.

In general, in translating $\exists u A[u]$, the variable u may be of arbitrary type τ and the translation of $A[u]$ may already start with some existential quantifiers, $A[u]^{\wedge} \equiv \exists v \forall w A_{\wedge}[u, v, w]$. In this general situation, we have to follow Schulte [35] in letting the bound X accompany all variables U, V in the existential part of the prefix of $(\exists u A[u])^{\wedge}$ as in def. 3.3.3. The special case above, however, is compatible with the general clause $(\exists)^{\wedge}$:

3.3.6   *Lemma*

The addition of the following clause to definition 3.3.3 does not extend the concept of $\wedge$-interpretability:

$(\exists^o)^\wedge$ $(\exists x^o A[x])^\wedge \equiv \exists X (X \# 0 \wedge \forall x \in X\, A[x])$ for $A[x] \in L(T_\in)$

Proof. 1. If X satisfies $X \# 0 \wedge \forall x \in \dot{X} A[x]$, then X and $Y = \lambda x.x$ satisfy $X \# 0 \wedge \forall x \in X\, A[Yx]$. Conversely, if X, Y satisfy $X \# 0 \wedge \forall x \in X\, A[Yx]$, then $X' = \{Yx \mid x \in X\}$ satisfies $X' \# 0 \wedge \forall x \in X'\, A[x]$. Hence, though not the same terms $\wedge$-interpret $\exists x^o A[x]$ according to $(\exists^o)^\wedge$ and to $(\exists)^\wedge$, $\wedge$-interpretability is preserved.
2. For longer formulae that contain $\exists x^o A[x]$ as subformula, the lemma follows, once the laws of intuitionistic logic have been $\wedge$-interpreted.

As remarked above, every existential variable in the prefix of $A^\wedge$ has an accompanying bound, inhabited or not. Therefore, if an additional inhabited bound is added in front of $A_\wedge$, it may, after distribution over the subformulae of $A_\wedge$, be contracted with the bounds already present and be finally absorbed altogether:

3.3.7   *Lemma: Absorption of bounds for the $\wedge$-translation*

Let $A^\wedge$ be $\exists v \forall w A_\wedge[v, w]$. Given terms S, V (not containing variables from w) there are terms v_0 (also not containing variables from w) such that

$$T_\in \vdash S \# 0 \wedge \forall s \in S\, A_\wedge[Vs, w] \to A_\wedge[v_0, w]$$

Proof by induction on A. We use the notation of the recursive definition 3.2.4 and 3.3.3. We also write $\forall s \in S \# 0$ shorthand for $S \# 0 \wedge \forall s \in S$.

$(T_\in)^\wedge$ $A^\wedge \equiv A \equiv A_\wedge$, v is empty. Hence v_0 is empty, and we have, as required, $T_\in \vdash S \# 0 \wedge \forall s \in S\, A \to A$.

$(\wedge)^\wedge$ We are given S, V, Y and argue in $T_\in$: the antecedent
$\forall s \in S \# 0\ (A_\wedge[Vs, w] \wedge B_\wedge[Ys, z])$ implies
$\forall s \in S \# 0\ A_\wedge[Vs, w] \wedge \forall s \in S \# 0\, B_\wedge[Ys, z]$.
By I.H., there are v_0 and y_0 (disjoint from w, z) for which this implies
$A_\wedge[v_0, w] \wedge B_\wedge[y_0, z]$ as required.

$(\to)^\wedge$ We are given S, X, W, Y and the antecedent
$\forall s \in S \# 0 (\forall x \in X svz\, A_\wedge[v, Wsxvz] \to B_\wedge[Ysv, z])$
which by distribution of $\forall s \in S$ implies
$\forall s \in S\, \forall x \in X svz\, A_\wedge[v, Wsxvz] \to \forall s \in S \# 0\, B_\wedge[Ysv, z]$

By I.H. for B, there is a Y_0 for the given S, Y for which $T_\in$ proves
$\forall s \in S\#0 \, B_\wedge[Ysv, z] \to B_\wedge[Y_0 v, z]$. Now we put
$X_0 vz := \{< s, x >| \, x \in Xsvz, s \in S\}$ and $W_0 x := W(x)_0(x)_1$,
and by contraction of bounds 3.1.12, we obtain from the last two lines
$\forall x \in X_0 vz A_\wedge[v, W_0 xvz] \to B_\wedge[Y_0 v, z]$.
This proves the implication from the first to the last line, which was to be shown.

$(\forall)^\wedge$ We are given terms S, V not containing variables from u, w. By I.H.
for $A[u]$, for the terms $S, V_1 := \lambda s.Vsu$, there is a tuple v_1 of terms not containing variables from w such that
$T_\in \vdash \forall s \in S\#0 A_\wedge[u, V_1 s, w] \to A_\wedge[u, v_1, w]$
Putting $v_0 := \lambda u.v_1$, this becomes, as desired,
$T_\in \vdash \forall s \in S\#0 A_\wedge[u, Vsu, w] \to A_\wedge[u, v_0 u, w]$

$(\forall \in)^\wedge$ is similar.

$(\exists)^\wedge$ Given terms S, X, U, V, we put
$X_0 := \{< s, x >| \, x \in Xs, s \in S\}$, $u_0 x := U(x)_0(x)_1$, $v_0 x := V(x)_0(x)_1$
and obtain by contraction of bounds 3.1.12 (without using the I.H.), as desired
$T_\in \vdash \forall s \in S\#0 \forall x \in Xs\#0 A_\wedge[Usx, Vsx, w] \to \forall x \in X_0\#0 A_\wedge[u_0 x, v_0 x, w]$

By subformula induction, lemma 3.3.7 follows. This absorption lemma extends the distribution lemmas by Burr and by Schulte from implications to the full language of $CZF^{\omega-}$. It is central for making the $\wedge$-translation work.

3.3.8 $\wedge$-*Interpretation Theorem for* $CZF^{\omega-}$

$CZF^{\omega-}$ *is* $\wedge$-*interpretable in* $T_\in$:

$$CZF^{\omega-} \overset{\wedge}{\hookrightarrow} T_\in$$

Proof by induction on deductions. The induction steps concerning the negative fragment, including equality and (T IND), or the axioms and rules of $T_\in$, including $A8, Q2b$ and $(T - EXT)$, are the same as in the proof of the $\wedge$-interpretation of $KP\omega^\omega$ 3.2.7. The axioms and rules that still have to be interpreted are therefore $A9, Q2, Q4b, Q4$, and, of course, $(StrongColl)$.

A9. $A_0 \to B, A_1 \to B \vdash A_0 \vee A_1 \to B$
Here, A_0, A_1 are Δ_0, and let $B^\wedge \equiv \exists y \forall z \, B_\wedge[y, z]$. By I.H. and case distinction 3.1.6, there are terms Y such that for $i \in 2$
$$T_\in \vdash A_i \to B_\wedge[Yi, z]$$

By Δ_0-separation, we define a bound $S := \{x \in 1 \mid A_0\} \cup \{x \in \{1\} \mid A_1\}$. For this $S \subset 2$, $T_\in \vdash A_0 \vee A_1 \to S\#0$ and $T_\in \vdash \forall s \in S\, B_\wedge[Ys, z]$, hence

$$T_\in \vdash A_0 \vee A_1 \to \forall s \in S\#0\, B_\wedge[Ys, z]$$

By absorption of bounds 3.3.7, there are terms y_0 such that

$$T_\in \vdash A_0 \vee A_1 \to B_\wedge[y_0, z] \;:$$

a combination of Δ_0-separation and absorption of bounds gives us the term tuple y_0 interpreting $A_0 \vee A_1 \to B$.
We now turn to the laws governing existential quantification.

Q2. $A[b] \to \exists u A[u]$
Let $A[u]^\wedge$ be $\exists v \forall w A_\wedge[u, v, w]$, and let $(\exists u A[u])^\wedge$ be $\exists X u y \forall z (\forall x \in X\#0\, A_\wedge[ux, yx, z])$. Then $(Q2)^\wedge$ is

$$\exists SWXUY \forall vz (\forall s \in Svz\, A_\wedge[b, v, Wsvz] \to \forall x \in X\#0\, A_\wedge[Uvx, Yvx, z])$$

Obviously, $Svz = X = 1, Wsvz = z, Uvx = b, Yvx = v$ $\wedge$-interpret Q2.

Q4b. $x \in t \wedge A[x] \to B \vdash \exists x \in t\, A[x] \to B$ for x not free in B.
This case is analogous to A9. Again, $A[x]$ is Δ_0. By I.H. and λ-abstraction, there are terms Y such that

$$T_\in \vdash x \in t \wedge A[x] \to B_\wedge[Yx, z]$$

By Δ_0-separation, we define $S := \{x \in t \mid A[x]\}$. Then, as above,

$$T_\in \vdash \exists x \in t\, A[x] \to S\#0 \wedge \forall s \in S\, B_\wedge[Ys, z]$$

and by absorption of bounds, we obtain y_0 interpreting $\exists x \in t\, A[x] \to B$.

Q4. $A[u] \to B \vdash \exists u\, A[u] \to B$ for u not free in B.
Let $A[u]^\wedge$ be as in Q2 and $B^\wedge$ as above. By I.H., also $\forall u(A[u] \to B)$ is $\wedge$-interpretable, i.e. there are terms X_0, W_0, Y_0 s.t.
(1) $T_\in \vdash \forall x_0 \in X_0 uvz\, A_\wedge[u, v, W_0 x_0 uvz] \to B_\wedge[Y_0 uv, z]$.
Let $(\exists u A[u])^\wedge := \exists S u'v' \forall w (\forall s \in S\#0\, A_\wedge[u's, v's, w])$. Then we are looking for terms X, W, Y with
(2) $T_\in \vdash \forall x \in XSu'v'z \forall s \in S\#0\, A_\wedge[u's, v's, WxSu'v'z] \to B_\wedge[YSu'v', z]$.
Here we pay a price for the additional bound S that, via the $\wedge$-translation, enters as a free variable into (2). Below, we shall refer to the antecedent of this formula as (2.l) and to its consequent as (2.r). In (1), we substitute $u's, v's$ for u, v and distribute the bound $S\#0 \wedge \forall s \in S$ over the implication (1). That yields
(3) $T_\in \vdash (3.l) \to (3.r)$ where
(3.l) $\forall s \in S\#0\; \forall x_0 \in X_0(u's)(v's)z\, A_\wedge[u's, v's, W_0 x_0(u's)(v's)z]$ and

$(3.r)$ $\forall s \in S\#0\ B_\wedge[Y_0(u's)(v's), z]$

Absorption of bounds 3.3.7 gives us terms $y_0[S, u', v']$ and
$Y := \lambda Su'v'.y_0[S, u', v']$ with $T_\in \vdash (3.r) \to (2.r)$.
It remains to show $T_\in \vdash (2.l) \to (3.l)$ for suitable X, W. In $(3.l)$, the
dependence of X_0 on $s(\in S)$ seems irritating. We make it disappear by
coding
$XSu'v'z := \{< x_0, s' >|\ x_0 \in X_0(u's')(v's')z, s' \in S\} \cup 1$ and
$WxSu'v'z := W_0(x)_0(u'(x)_1)(v'(x)_1)z$
Due to the added $\cup 1$, $XSu'v'z$ is inhabited, and we may interchange
$\forall x \in XSu'v'z$ and $\forall s \in S\#0$. Then we see that, in $T_\in$, $(2.l)$ implies

$$\forall s \in S\#0\,\forall s' \in S\,\forall x_0 \in X_0(u's')(v's')z\ A_\wedge[u's, v's, W_0x_0(u's')(v's')z]$$

which, by cutting back the product $S \times S$ of the first two bounds to its
diagonal $\{< s, s' >|\ s \in S\}$, implies $(3.l)$. The now completed chain of
implications yields (2) for the terms X, W, Y as presented.

From this point on, the proof of lemma 3.3.6 is completed. Its application
simplifies matters slightly in the sequel.

$(StrongColl)$ $\forall x \in a \exists y\ F[x, y] \to \exists b\ F'[a, b]$ where

$$F'[a, b] :\equiv \forall x \in a \exists y \in b\ F[x, y] \wedge \forall y \in b \exists x \in a\ F[x, y]$$

Case 1. $F[x, y]$ is Δ_0. In this case, $F'[a, b]$ is also Δ_0. Applying lemma 3.3.6
to the quantifier $\exists y$, we obtain
$(\forall x \in a \exists y\ F[x, y])^\wedge \equiv \exists Y \forall x \in a\ (Yx\#0 \wedge \forall y \in Yx\ F[x, y])$
We then define $b[Y]$ outright, i.e. setting its accompanying bound $= 1$, by

$$b := \mathsf{U}Ya = \bigcup\{Yx \mid x \in a\}$$

The antecedent $\forall x \in a\ (Yx\#0 \wedge \forall y \in Yx\ F[x, y])$ implies
$\forall x \in a \exists y \in \mathsf{U}Ya\ F[x, y]$. Moreover, $\forall y \in \mathsf{U}Ya \exists x \in a\ y \in Yx$. Hence the
antecedent also implies $\forall y \in \mathsf{U}Ya \exists x \in a\ F[x, y]$, and we get

$$T_\in \vdash \forall x \in a\ Yx\#0 \wedge \forall y \in Yx\ F[x, y] \to F'[a, \mathsf{U}Ya],$$

the desired $\wedge$-interpretation of Δ_0-$(StrongColl)$.
So it is just the clause $(\exists^o)^\wedge$ which in this case produces the smooth
interpreting term $b = \mathsf{U}Ya$. Looking at it the other way, $(StrongColl)$ more
or less forces us to translate the existential quantifier as we do.

Case 2. $F[x, y]$ is not Δ_0, $F[x, y]^\wedge \equiv \exists v \forall w F_\wedge[x, y, v, w]$. Then
$(\forall x \in a \exists y F[x, y])^\wedge \equiv \exists SY v \forall w (\forall x \in a \forall s \in Sx\#0\ F_\wedge[x, Yxs, vxs, w])$

and, again suppressing the bound accompanying b, we have

$$(\exists b F'[a,b])^{\wedge} \equiv \exists b S_0 Y_0 v_0 S_1 X_1 v_1 \forall w_0 w_1$$

$$\left(\forall x \in a \forall s \in S_0 x \# 0 \, (Y_0 x s \in b \wedge F_{\wedge}[x, Y_0 x s, v_0 x s, w_0]\right)$$

$$\wedge \, \forall y \in b \forall s \in S_1 y \# 0 \, (X_1 y s \in a \wedge F_{\wedge}[X_1 y s, y, v_1 y s, w_1]))$$

and finally, abbreviating $tSYvw_0w_1$ by $t+$ and $tSYv$ by $t-$, we get

$(StrongColl)^{\wedge} \equiv$

$$\exists Z W b S_0 Y_0 v_0 S_1 X_1 v_1 \, \forall SY vw_0 w_1$$

$$\left(\forall i \in Z + \forall x \in a \forall s \in Sx \# 0 \, F_{\wedge}[x, Yxs, vxs, Wi+]\right.$$

$$\rightarrow \left(\forall x \in a \forall s \in S_0 x - \# 0 \, (Y_0 x s - \in b - \wedge F_{\wedge}[x, Y_0 x s -, v_0 x s -, w_0]\right)$$

$$\wedge \, \forall y \in b - \forall s \in S_1 y - \# 0 \, (X_1 y s - \in a \wedge F_{\wedge}[X_1 y s -, y, v_1 y s -, w_1]))\big)$$

As the antecedent is to imply a conjunction of two conjuncts, we start by setting $Z+ := 2$ and $Wi+ := w_i$ for $i \in 2 = \{0, 1\}$. Then the first conjunct easily follows from the antecedent for $i = 0$, if we put

$b- := \{Yxs \mid s \in Sx, x \in a\}$ (which clearly corresponds to the choice of b in case 1), $S_0 x- := Sx, Y_0 x s- := Yxs, v_0 x s- := vxs$. To make the antecedent for $i = 1$ imply the second conjunct, we have to define, for any $y \in b-$, the set of those x for which y is an element Yxs. Following Schulte [35], we put

$$S_1 y- := \{< x, s > \mid y = Yxs, s \in Sx, x \in a\}$$
$$= \{\mathsf{K} < x, s > z \mid z \in \{y\} \cap \{Yxs\}, s \in Sx, x \in a\}$$

which is an (iterated) application of operation (8) in 3.1.2. We then have

$$T_{\in} \vdash \forall y \in b- \, S_1 y - \# 0,$$

and for $X_1 y s- := (s)_0$ and $v_1 y s- := v(s)_0(s)_1$, the antecedent also implies the second conjunct.

This completes the proof of the $\wedge$-interpretation theorem. Since $\wedge$ is the identity on $L(T_{\in})$, it immediately yields:

3.3.9   *Corollary: Conservativity and relative consistency*

$CZF^{\omega-}$ is a conservative extension of $T_{\in}$. The consistency of $T_{\in}$ implies the consistency of $CZF^{\omega-}$.

As the $\wedge$-translation is not idempotent, the $\wedge$-interpretation theorem does not extend immediately to $CZF^{\omega-} + \{A \leftrightarrow A^{\wedge}\}$ by lemma 0.1.19 (2). We postpone this question to the end of the section.

As in arithmetic, it seems adequate also in constructive set theory to formulate a Markov rule of which the conclusion is the $\wedge$-translation of the premise. Closure of $CZF^{\omega-}$ under this rule is then also an immediate corollary:

3.3.10 Proposition: Closure under Markov rule

(Rule-$M_\in$) *is admissible in* $CZF^{\omega-}$:

(Rule-$M_\in$) *If* $\vdash \forall w A[w] \to B$ *with* $A, B \in L(T_\in)$ *then*

$$\vdash \exists X W (\forall x \in X\ A[Wx] \to B)$$

Admissibility of further rules will be studied in the next section.
As a side remark, we mention a further analogy to the arithmetic case.

3.3.11 The type o version of $CZF^{\omega-}$

The type o fragment $CZF_0^{\omega-}$ of $CZF^{\omega-}$ is an extension of $T_{\in 0}$ in exactly
the same way as $CZF^{\omega-}$ is an extension of $T_\in$ (cf. 3.1.1(8)). If, in a formula
B of $L(CZF^{\omega-})$, we replace all equations $a = b$ of a higher type τ by the
formula $\forall Y^{\tau \to o} Ya = Yb$ of type o, we obtain a formula B^o of type o, and
we have $B^o \equiv B$ for B of type o. By induction on deductions, it is seen:

3.3.12 Proposition

If $CZF^{\omega-} \vdash A$, *then* $CZF_0^{\omega-} \vdash A^o$.
$CZF^{\omega-}$ *is a conservative extension of* $CZF_0^{\omega-}$.

The *proof* parallels the proof of proposition 1.3.15.

Since the $\wedge$-translation of a formula of type o is again a formula of type o,
we get as a corollary to our interpretation theorem:

3.3.13 Corollary: $\wedge$-interpretation of $CZF_0^{\omega-}$ and CZF^-

$$CZF_0^{\omega-} \overset{\wedge}{\hookrightarrow} T_{\in 0} \quad and \quad CZF^- \overset{\wedge}{\hookrightarrow} T_{\in 0}$$

Since $T_\in$ is a subtheory of $CZF^{\omega-}$, it may also be subjected to the
^{o}-translation. Putting our conservativity results together, we have:

3.3.14 Corollary

$CZF^{\omega-}, CZF_0^{\omega-}, T_\in$ *are all conservative extensions of* $T_{\in 0}$.

Over $CZF^{\omega-}$, the strength of the $\wedge$-translation, the schema $\{A \leftrightarrow A^\wedge\}$
is characterized by schemata quite similar to those characterizing the $\wedge$-
translation over HA^ω.

3.3.15   The theory $CZF^{\omega+}$

$CZF^{\omega+}$ is the extension of $CZF^{\omega-}$ by a **schema of weak independence of premise** $WIP_\in$, a **Markov schema** $M_\in$, and a **schema of weak choice** WAC as follows:

$WIP_\in$ $(\forall w A \to \exists y B[y]) \to \exists XY (\forall w A \to X \# 0 \land \forall x \in X\ B[Yx])$
 for $A \in L(T_\in)$

$M_\in$ $(\forall w A[w] \to B) \to \exists XW (\forall x \in X\ A[Wx] \to B)$ for $A, B \in L(T_\in)$

WAC $\forall x \exists y A[x,y] \to \exists SY \forall x (Sx \# 0 \land \forall s \in Sx\ A[x, Ysx])$

Here, w, y, and in WAC also x are to be read as non-empty tuples of variables of arbitrary type. Bounds X, Sx are single variables resp. terms of type o.

$$CZF^{\omega+} :\equiv CZF^{\omega-} + WIP_\in + M_\in + WAC$$

The Markov principle $M_\in$ is the literal analog of the Markov principle $M_\land$ discussed in connection with the $\land$-translation of HA^ω which in turn is a weakened version of the standard Markov principle M, though extended to a language with undecidable atomic formulae. $WIP_\in$ is weaker than corresponding arithmetical principles of independence of premise: Given the antecedent $\forall w\ A \to \exists y\ B[y]$, it does not produce a y satisfying $\forall w\ A \to B[y]$, independent of the information $\forall w\ A$, but it only produces X-many such objects, where the inhabitedness of X depends on the information $\forall w\ A$. Similarly, WAC is not really a choice principle: Given the antecedent $\forall x \exists y\ A[x,y]$, it does not produce a choice function, but rather a choice relation, where the codomain $\{Ysx \mid s \in Sx\}$ varies with x. The weak formulations of $WIP_\in$ and WAC correspond to the weak translation $(\exists)^\land$ of the existential quantifier in set theory.

3.3.16   Lemma

In $CZF^{\omega+}$ is provable:

1. $(A^\land \to B^\land) \quad \to (A \to B)^\land$
2. $\forall x \in t(A[x]^\land) \to (\forall x \in t\ A[x])^\land$
3. $\forall u(A[u]^\land) \qquad \to (\forall u A[u])^\land$

Proof. 1. $A^\land \to B^\land \equiv \exists v \forall w A_\land \to \exists y \forall z B_\land$ is logically equivalent to $\forall v(\forall w A_\land \to \exists y \forall z B_\land[y,z])$. By $WIP_\in$, this implies $\forall v \exists XY (\forall w A_\land \to X \# 0 \land \forall x \in X \forall z B_\land[Yx, z])$ which, in turn, is logically equivalent to

$\forall v \exists XY \forall z (\forall w A_\wedge \to X \# 0 \wedge \forall x \in X \ B_\wedge[Yx,z])$
By absorption of bounds 3.3.7, applied to bound X and the consequent, this is equivalent to
$\forall v \exists y \forall z (\forall w A_\wedge \to B_\wedge[y,z])$. By $M_\in$, this implies
$\forall v \exists y \forall z \exists XW (\forall x \in XA_\wedge[v,Wx] \to B_\wedge[y,z])$
which by WAC implies
$\exists SXWY \forall vz (Svz \# 0 \wedge \forall s \in Svz (\forall x \in XsvzA_\wedge[v,Wsxvz] \to B_\wedge[Ysv,z]))$.
Again by absorption of bounds, applied to bound S and the whole formula, this is finally equivalent to
$\exists XWY \forall vz (\forall x \in XvzA_\wedge[v,Wxvz] \to B_\wedge[Yv,z]) \equiv (A \to B)^\wedge$
3. $\forall u (A[u]^\wedge) \equiv \forall u \exists v \forall w A_\wedge[u,v,w]$ implies by WAC
$\exists SV \forall u \forall s \in Su \# 0 \forall w \ A_\wedge[u,Vsu,w]$ which logically implies
$\exists SV \forall u \forall w \forall s \in Su \# 0 \ A_\wedge[u,Vsu,w]$.
By absorption of bounds 3.3.7, this implies
$\exists v_0 \forall uw A_\wedge[u,v_0u,w] \equiv (\forall u A[u])^\wedge$.
2. follows from 1. and 3.

3.3.17 Characterization Theorem

$$CZF^{\omega+} \equiv CZF^{\omega-} + \{A \leftrightarrow A^\wedge\}$$

The *proof from left to right* is by induction on A. The induction steps that are proved already in $CZF^{\omega-}$ are listed in lemma 3.3.4. The other induction steps that need additional axioms from $CZF^{\omega+}$ are listed in lemma 3.3.16.

Proof from right to left. For a derivation of $WIP_\in$, we exploit the schema $\{A \leftrightarrow A^\wedge\}$ twice:
The $\wedge$-translation of the antecedent $\forall w \ A[w] \to \exists y \ B[y]$ of $WIP_\in$ is
$\exists SWXYY' \forall z (\forall s \in Sz \ A[Ws] \to \forall x \in X \# 0 B_\wedge[Yx,Y'x,z])$,
since A is in $L(T_\in)$, which implies
$\exists XYY' \forall z (\forall w A \to \forall x \in X \# 0 \ B_\wedge[Yx,Y'x,z])$
which in turn implies
$\exists XY (\forall w A \to \forall x \in X \# 0 \exists y' \forall z B_\wedge[Yx,y',z])$.
Since $\exists y' \forall z B_\wedge[Yx,y',z])$ is $B[Yx]^\wedge$, $WIP_\in$ follows.

The consequent of $M_\in$ is the $\wedge$-translation of its antecedent. Hence $M_\in$ is an axiom $C \to C^\wedge$.

For WAC, we proceed as for $WIP_\in$. The $\wedge$-translation of the antecedent is
$\exists SYV \forall xw \forall s \in Sx \# 0 \ A_\wedge[x,Ysx,Vsx,w]$ which implies
$\exists SY \forall x \forall s \in Sx \# 0 \exists v \forall w A_\wedge[x,Ysx,v,w]$.
Since $\exists v \forall w A_\wedge[x,Ysx,v,w]$ is $A[x,Ysx]^\wedge$, WAC follows.

Due to the characterization theorem, it can be shown that $CZF^{\omega+}$ is not only $\wedge$-interpretable in $T_\in$, but also $\wedge$-maximal relative to $T_\in$.

3.3.18   Lemma

If $CZF^{\omega-} \vdash A^\wedge \to B^\wedge$, then $A \to B$ is $\wedge$-interpretable in $T_\in$.

Proof. Let $CZF^{\omega-} \vdash \exists v \forall w A \to \exists y \forall z B$. By $\wedge$-interpretation, there are terms X, W, S', Y such that for free variables S, V, z,
$$T_\in \vdash \forall x \in XSVz \, \forall s \in S\#0 \, A_\wedge[Vs, WxSVz] \to \forall s \in S'SV\#0 \, B_\wedge[YssV, z]$$
First, we substitute 1 for S and $\lambda s.v$ for V. Then we set $X_0vz := X1(\lambda s.v)z$ and $W_0xvz := Wx1(\lambda s.v)z$, and obtain
$$T_\in \vdash \forall x \in X_0vz A_\wedge[v, W_0xvz] \to \forall s \in S'1(\lambda s.v)\#0 B_\wedge[Ys1(\lambda s.v), z]$$
Absorption of bounds 3.3.7 simplifies the consequent of this implication to $B_\wedge[Y_0v, z]$ for suitable Y_0. So, X_0, W_0, Y_0 $\wedge$-interpret $A \to B$ in $T_\in$.

3.3.19   Extended $\wedge$-Interpretation Theorem

$$CZF^{\omega+} \equiv CZF^{\omega-} + \{A \leftrightarrow A^\wedge\} \overset{\wedge}{\hookrightarrow} T_\in$$
and both these theories are $\wedge$-maximal for $T_\in$.

Proof. By lemma 3.3.4.7, $\vdash A^\wedge \leftrightarrow A^{\wedge\wedge}$. Hence, by the last lemma, $A \leftrightarrow A^\wedge$ is $\wedge$-interpretable in $T_\in$. Therefore, by the $\wedge$-interpretation theorem 3.3.8 and by lemma 0.1.19 (1), $CZF^{\omega-} + \{A \leftrightarrow A^\wedge\}$ is $\wedge$-interpretable in $T_\in$ and $\wedge$-maximal for $T_\in$. By the characterization theorem 3.3.17, the same holds for $CZF^{\omega+}$.

3.3.20   Corollary

$CZF^{\omega+}$ is a conservative extension of $T_\in$. The consistency of $T_\in$ implies the consistency of $CZF^{\omega+}$.

3.4   Modified realizations of constructive set theory

Technically simpler functional interpretations than the $\wedge$-interpretation of $CZF^{\omega-}$ in the previous section are modified realization mr of $CZF^{\omega-}$ and its hybrids mrt and mq. They have first been considered by Schulte [35]. We proceed in analogy to section 1.6, writing again $v \; mr \; A$ for $A_{mr}[v]$, etc. In this section, we consider the bounded universal quantifier $\forall x \in t$ a primitive connective only within the language of $T_\in$.

3.4.1 Recursive definitions of modified realization translations mr, mrt, mq on $L(CZF^{\omega -})$

Let I equally refer to mr, mrt, and mq.

$(T_\in)^I \qquad A^I \equiv A$ for $A \in L(T_\in)$

Let $A^I \equiv \exists v \; v \, I \, A$ and $B^I \equiv \exists y \; y \, I \, B$. The recursive clauses $(\wedge)^I$ for conjunction, $(\to)^I$ for implication, and $(\forall)^I$ for universal quantification coincide with the corresponding clauses in definition 1.6.1. In clauses $(\wedge)^I$ and $(\to)^I$, it is again assumed that not both components A, B are from $L(T_\in)$.

$(\exists)^{mr} \quad (\exists u A[u])^{mr} \equiv \exists XUV \; X \# 0 \wedge \forall x \in X \; Vx \, mr \, A[Ux]$

$(\exists)^{mrt} \quad (\exists u A[u])^{mrt} \equiv \exists XUV \; X \# 0 \wedge \forall x \in X \; Vx \, mrt \, A[Ux]$

$(\exists)^{mq} \quad (\exists u A[u])^{mq} \equiv \exists XUV \; X \# 0 \wedge \forall x \in X \; (A[Ux] \wedge Vx \, mq \, A[Ux])$

The only difference to definition 1.6.1 lies in the clauses $(\exists)^I$ for $I = mr, mrt, mq$: Here, I, like $\wedge$, does not call for an explicit instance, but only for an explicit inhabited set of objects satisfying $A[u]$. Therefore, mr is not idempotent, as it is in arithmetic.

The same close relation as in arithmetic holds between mrt and mq:

3.4.2 Proposition

$$CZF^{\omega -} \vdash v \, mrt \, A \leftrightarrow (A \wedge v \, mq \, A)$$

The *proof* by subformula induction is identical with the proof of proposition 1.6.2, with the exception of the induction step for the existential quantifier:

$(\exists) \quad X, U, V \, mrt \, \exists u A[u] \equiv \forall x \in X \# 0 \; Vx \, mrt \, A[Ux]$

is by I.H. equivalent to $\forall x \in X \# 0 \; (A[Ux] \wedge Vx \, mq \, A[Ux])$.

Since $\forall x \in X \# 0 \; A[Ux]$ implies $\exists u A[u]$, this is equivalent to

$$\exists u A[u] \wedge \forall x \in X \# 0 \; (A[Ux] \wedge Vx \, mq \, A[Ux]) \equiv \exists u A[u] \wedge X, U, V \, mq \, \exists u A[u]$$

Thus, as in arithmetic, mrt-realizing terms are automatically mq-realizing terms, and we again obtain a unified treatment of mr, mrt, mq by introducing mrG-realization for an arbitrary sentence G of $L(CZF^{\omega -})$ in analogy to definition 1.6.3. For definiteness, we give this definition in full detail.

3.4.3 Recursive definition of realization translations mrG

Let G be a sentence of $L(CZF^{\omega -})$. To any formula A of $L(CZF^{\omega -})$, mrG assigns a formula $A^{mrG} \equiv \exists v \; v \, mrG \, A$ of $L(CZF^{\omega -})$ as follows:

$(T_\in)^{mrG} \qquad A^{mrG} \equiv A \quad$ for $A \in L(T_\in)$

Let A^{mrG} be as above and $B^{mrG} \equiv \exists y \; y \; mrG \; B$; then

$(\wedge)^{mrG} \quad (A \wedge B)^{mrG} \equiv \exists vy(v \; mrG \; A \wedge y \; mrG \; B)$

$(\rightarrow)^{mrG} \quad (A \rightarrow B)^{mrG} \equiv \exists Y((G \rightarrow A \rightarrow B) \wedge \forall v(v \; mrG \; A \rightarrow Yv \; mrG \; B))$

$(\forall)^{mrG} \quad (\forall u A[u])^{mrG} \equiv \exists V \forall u(Vu \; mrG \; A[u])$

$(\exists)^{mrG} \quad (\exists u A[u])^{mrG} \equiv \exists XUV \; (X \# 0 \wedge \forall x \in X \; Vx \; mrG \; A[Ux])$

In spite of the essential difference in the definition of $(\exists)^{mrG}$ to the arithmetic case, the *formal* equivalence still holds, even for tuples of variables u, as it does for the translation $\wedge$:

$$(\exists u A[u])^{mrG} \leftrightarrow \exists uv(v \; mrG \; A[u])$$

Interesting for us are again the cases $G \equiv \bot$ and $G \equiv \top$, because

$$mr\bot \equiv mr \quad \text{and} \quad mr\top \equiv mrt$$

For $\exists$free and Harrop formulae, arguments transfer from the arithmetic case in section 1.6 to the present situation. Their definition is adapted to the language of set theory as follows:

3.4.4   *Definition: $\exists$free and Harrop formulae*

A formula A of $L(CZF^{\omega -})$ is $\exists$**free**, if A is a formula of $L(KP\omega^{\omega})$, i.e. if $\exists$ occurs in A at most in Δ_0-subformulae of A. **Harrop formulae** are, as in arithmetic, formulae without strictly positive occurrences of unbounded $\exists$; their recursive definition therefore runs as in 1.6.13, with $L(T)$ replaced by $L(T_\in)$ in the starting clause.

3.4.5   *Lemma*

(1) *For I any of the above translations (including mq) and A Harrop,*

$$A^I \equiv \emptyset \; I \; A$$

(2) *For these translations I and $\exists$free A, $CZF^{\omega -} \vdash A^I \leftrightarrow A$, and $A^{mr} \equiv A$.*

(3) *For Harrop formulae A, $CZF^{\omega -} \vdash G \rightarrow A \rightarrow \emptyset \; mrG \; A$*

(4) *For all formulae A, $v \; mr \; A$ is $\exists$free.*

Further basic properties of mrG transfer from the arithmetic case:

(5) $CZF^{\omega -} \vdash G \rightarrow (v \; mrG \; A \leftrightarrow v \; mrt \; A)$

(6) $CZF^{\omega -} \vdash v \; mrG \; A \rightarrow (G \rightarrow A)$

(7) *If in an extension Th of $CZF^{\omega -}$*

$$Th \vdash (G \rightarrow C \rightarrow D) \wedge (v \; mrG \; C \rightarrow 1, \lambda s.v \; mrG \; D)$$

then the tuple $\lambda v.1, \lambda vs.v \quad mrG$-realizes $(C \rightarrow D)$ in Th.

For *proofs*, cf. lemmata 1.6.4 to 1.6.6.

3.4.6 Lemma

The addition of the following clause to definition 3.4.3 does not extend the concept of mrG-realizability:

$$(\exists^o)^{mrG} \quad (\exists x^o A[x])^{mrG} \equiv \exists X (X \# 0 \wedge \forall x \in X\ A[x]) \quad for \quad A[x] \in L(T_\in)$$

The *proof* may be taken over from lemma 3.3.6, replacing the translation $\wedge$ by mrG.

Variables in the prefix of A^{mrG} have inhabited accompanying bounds. An absorption lemma for additional inhabited bounds, similar to lemma 3.3.7, holds also for mrG:

3.4.7 Lemma: Absorption of bounds for mrG

Let $A^{mrG} \equiv \exists v\ v\ mrG\ A$. Given terms $S : o, V$, there are terms v_0 such that

$$CZF^{\omega^-} \vdash S \# 0 \wedge \forall s \in S\ Vs\ mrG\ A \to v_0\ mrG\ A$$

The *proof* by subformula induction is, with the exception of the induction step for implication, identical with the special case of the proof of lemma 3.3.7 where the universal part of the prefix of $A^\wedge$ is empty.

$$(\to) \qquad \forall s \in S \# 0\ Ys\ mrG\ (A \to B) \equiv$$
$$\forall s \in S \# 0 (G \to A \to B) \wedge \forall v(v\ mrG\ A \to Ysv\ mrG\ B)$$

which implies
$(G \to A \to B) \wedge \forall v(v\ mrG\ A \to \forall s \in S \# 0\ Ysv\ mrG\ B)$
which by I.H. for B implies for some terms Y_0
$(G \to A \to B) \wedge \forall v(v\ mrG\ A \to Y_0 v\ mrG\ B) \equiv Y_0\ mrG\ (A \to B)$

3.4.8 Theorem: mrG-realization of CZF^{ω^-}

$$CZF^{\omega^-} \overset{mrG}{\hookrightarrow} CZF^{\omega^-}$$

More precisely: If $CZF^{\omega^-} \vdash A$, then there is a tuple b of terms with $FV(b) \subset FV(A)$ such that for all sentences G:

$$CZF^{\omega^-} \vdash b\ mrG\ A$$

Proof by induction on deductions. For the axioms and rules of BLTT, but also for the laws governing the universal quantifier Q1, Q3, the proof may be taken over from the proof of the modified realization theorem 1.6.7, because that proof does not make use of any arithmetical properties. For A8, Q1b to Q3b, for axioms of $T_\in$ defining set theoretic constants and also for $(T - EXT)$, there is nothing to prove. For the remaining axioms and rules, we cut the $\wedge$-interpretation in theorem 3.3.8 down to mrG-realization of $CZF^{\omega-}$. We mostly suppress the G-part and concentrate on the mr-part of the statements to be proved.

A9. $A_0 \to B, A_1 \to B \vdash A_0 \vee A_1 \to B$
A_0, A_1 are Δ_0; let $B^{mrG} \equiv \exists y \; y \, mrG \, B$.
By I.H. and case distinction 3.1.6, there are terms Y such that for $i \in 2$

$$CZF^{\omega-} \vdash A_i \to Yi \, mrG \, B$$

By Δ_0-separation, define $S := \{x \in 1 \mid A_0\} \cup \{x \in \{1\} \mid A_1\}$. For this $S \subset 2$,

$$CZF^{\omega-} \vdash A_0 \vee A_1 \to \forall s \in S \# 0 \; Ys \, mrG \, B$$

By absorption of bounds 3.4.7, there are terms y_0 such that

$$CZF^{\omega-} \vdash A_0 \vee A_1 \to y_0 \, mrG \, B$$

So, $y_0 \, mrG$-realizes $A_0 \vee A_1 \to B$.

Q2. $A[b] \to \exists u A[u]$. Let $A[b]^{mrG} \equiv \exists v \; v \, mrG \, A[b]$, and
$(\exists u A[u])^{mrG} \equiv \exists X u y \, \forall x \in X \# 0 \; yx \, mrG \, A[ux]$. Then $(Q2)^{mrG}$ is

$$(G \to Q2) \wedge \exists XUY \forall v(v \, mrG \, A[b] \to \forall x \in Xv \# 0 \; Yvx \, mrG \, A[Uvx])$$

Thus, $Xv = 1, Uvx = b, Yvx = v \quad mrG$-realize Q2.

Q4b. $x \in t \wedge A[x] \to B \vdash \exists x \in t \, A[x] \to B$ for x not free in B.
$A[x]$ is Δ_0. By I.H. and λ-abstraction, there are terms Y s.t.

$$CZF^{\omega-} \vdash x \in t \wedge A[x] \to Yx \, mrG \, B$$

By Δ_0-separation, we define $S := \{x \in t \mid A[x]\}$. Then, as above,

$$CZF^{\omega-} \vdash \exists x \in t \, A[x] \to \forall s \in S \# 0 \; Ys \, mrG \, B$$

Again, by absorption of bounds 3.4.7, there are terms y_0 for which the consequent implies $y_0 \, mrG \, B$. These $y_0 \, mrG$-realize $\exists x \in t \, A[x] \to B$.

Q4. $A[u] \to B \vdash \exists u \, A[u] \to B$ for u not free in B.
By I.H. and λ-abstraction, there are terms Y_0 such that
(1) $CZF^{\omega-} \vdash \forall v(v \, mrG \, A[u] \to Y_0 uv \, mrG \, B)$.
Let $(\exists u A[u])^{mrG} \equiv \exists SUV \forall s \in S \# 0 \; Vs \, mrG \, A[Us]$.
Then we are looking for terms Y such that

(2) $CZF^{\omega-} \vdash \forall SUV (\forall s \in S\#0\ Vs\,mrG\,A[Us] \to YSUV\,mrG\,B)$.

In (1), we substitute Us, Vs for u, v and distribute $\forall s \in S\#0$ over the implication. That yields

(3) $\vdash \forall SUV (\forall s \in S\#0\ Vs\,mrG\,A[Us] \to \forall s \in S\#0\ Y_0(Us)(Vs)\,mrG\,B)$

By absorption of bounds, there are terms $y_0[S, U, V]$ such that

(4) $CZF^{\omega-} \vdash \forall s \in S\#0\ Y_0(Us)(Vs)\,mrG\,B \to y_0\,mrG\,B$.

Finally, $Y := \lambda SUV.y_0[S, U, V]$, by (3) and (4), satisfies (2) and thus mrG-realizes $Q4$.

From here onwards, we may apply lemma 3.4.6.

(T IND) $\forall u \in a\ F[u] \to F[a] \vdash F[t]$

By I.H., there are terms Y_0 such that

(1) $CZF^{\omega-} \vdash \forall u \in a\ vu\,mrG\,F[u] \to Y_0va\,mrG\,F[a]$

We define terms Y by simultaneous transfinite recursion 3.1.13

$$Ya = Y_0(Y \upharpoonright a)a,$$

substitute $Y \upharpoonright a$ for v in (1), and obtain, because $\forall u \in a\ (Y \upharpoonright a)u = Yu$,

$$CZF^{\omega-} \vdash \forall u \in a\ Yu\,mrG\,F[u] \to Ya\,mrG\,F[a]$$

So, by (T IND), $CZF^{\omega-} \vdash Yt\,mrG\,F[t]$ follows.

$(StrongColl)$ $\quad \forall x \in a \exists y\ F[x, y] \to \exists b\ F'[a, b]$ $\quad$ where

$$F'[a, b] :\equiv \forall x \in a \exists y \in b\ F[x, y] \wedge \forall y \in b \exists x \in a\ F[x, y]$$

Case 1. $F[x, y]$ is Δ_0. In this case, $F'[a, b]$ is also Δ_0. Applying lemma 3.4.6 to the quantifier $\exists y$, we obtain

$(\forall x \in a \exists y\ F[x, y])^{mrG} \equiv \exists Y \forall x \in a \forall y \in Yx\#0\ F[x, y]$

We then define $b[Y]$ outright, i.e. setting its accompanying bound $= 1$, by

$$b := \mathsf{U}Ya = \bigcup\{Yx \mid x \in a\}$$

$\forall x \in a \forall y \in Yx\#0\ F[x, y]$ implies $\forall x \in a \exists y \in \mathsf{U}Ya\ F[x, y]$. By definition of $\mathsf{U}Ya$, we have $\forall y \in \mathsf{U}Ya \exists x \in a\ y \in Yx$. Hence the antecedent also implies $\forall y \in \mathsf{U}Ya \exists x \in a\ F[x, y]$, and we get

$$T_\in \vdash \forall x \in a\ Yx\#0 \wedge \forall y \in Yx\ F[x, y] \to F'[a, \mathsf{U}Ya],$$

the desired mrG-interpretation of Δ_0-$(StrongColl)$.

So, also for modified realization, $(StrongColl)$ more or less forces us to translate the existential quantifier as we do.

Case 2. $F[x, y]$ is not Δ_0, $F[x, y]^{mrG} \equiv \exists v\ v\,mrG\,F[x, y]$. Then

(1) $(\forall x \in a \exists y F[x, y])^{mrG} \equiv \exists SYV (\forall x \in a \forall s \in Sx\#0\ Vxs\,mrG\,F[x, Yxs])$

and, again suppressing the bound accompanying b, we have

$\exists b \, F'[a,b])^{mrG} \equiv \exists b S_0 Y_0 V_0 S_1 X_1 V_1$

(2) $(\forall x \in a \forall s \in S_0 x \# 0 \, (Y_0 x s \in b \wedge V_0 x s \, mrG \, F[x, Y_0 x s])$

(3) $\wedge \, \forall y \in b \forall s \in S_1 y \# 0 \, (X_1 y s \in a \wedge V_1 y s \, mrG \, F[X_1 y s, y]))$

and finally, abbreviating $tSYv$ by $t-$,

$(StrongColl)^{mrG} \equiv (G \to (StrongColl)) \wedge \exists b S_0 Y_0 V_0 S_1 X_1 V_1 \forall SY v$

(1') $(\forall x \in a \forall s \in Sx \# 0 \, vxs \, mrG \, F[x, Y x s]) \to$

(2') $(\forall x \in a \forall s \in S_0 x - \# 0 (Y_0 x s - \in b - \wedge V_0 x s - \, mrG \, F[x, Y_0 x s-])$

(3') $\wedge \, \forall y \in b - \forall s \in S_1 y - \# 0 \, (X_1 y s - \in a \wedge V_1 y s - \, mrG \, F[X_1 y s-, y]))$

(1') implies (2'), if we put $bSYv := \{Y x s \mid s \in Sx, x \in a\}$
(which corresponds to the choice of b in case 1),
$S_0 x S Y v := Sx, \quad Y_0 x s S Y v := Y x s, \quad V_0 x s S Y v := v x s.$
To make (1') imply (3'), we have to define, for any $y \in bSYv$, the set of
those x for which y is an element $Y x s$. As for the $\wedge$-interpretation, we put

$$S_1 y S Y v := \{< x, s > \mid y = Y x s, s \in Sx, x \in a\}$$
$$= \{\mathsf{K} < x, s > z \mid z \in \{y\} \cap \{Y x s\}, s \in Sx, x \in a\}$$

which is an (iterated) application of operation (8) in 3.1.2. We then have

$$T_\in \vdash \forall y \in bSYv \quad S_1 y S Y v \# 0 \, ,$$

and for $X_1 y s S Y v := (s)_0$ and $V_1 y s S Y v := v(s)_0(s)_1$, (1') also implies (3').
This completes the proof of the mrG-realization theorem. Immediate
corollaries are:

3.4.9   *Uniform modified realization theorem*

*If $CZF^{\omega-} \vdash A$, then there is a tuple of terms b with $FV(b) \subset FV(A)$ such
that*

$$CZF^{\omega-} \vdash b \, mr \, A \wedge b \, mrt \, A \wedge b \, mq \, A$$

Hence

$$CZF^{\omega-} \overset{mr}{\hookrightarrow} CZF^{\omega-}, CZF^{\omega-} \overset{mrt}{\hookrightarrow} CZF^{\omega-}, \quad and \quad CZF^{\omega-} \overset{mq}{\hookrightarrow} CZF^{\omega-}$$

Proof. This follows from theorem 3.4.8 for mr and mrt, because $mr \equiv mr\bot$
and $mrt \equiv mr\top$, and for mq by proposition 3.4.2.

Though the regular realization translation mr is not idempotent, due to the
absorption lemma, mr extends beyond $CZF^{\omega-}$ to a mr-maximal extension:

3.4.10   Corollary

$$CZF^{\omega-} + \{A \leftrightarrow A^{mr}\} \overset{mr}{\hookrightarrow} CZF^{\omega-}$$

and $CZF^{\omega-} + \{A \leftrightarrow A^{mr}\}$ is mr-maximal relative to $CZF^{\omega-}$.

Proof. The second statement follows from the first by lemma 0.1.19. For the first, in view of theorem 3.4.8, it suffices to mr-realize $A \to A^{mr}$ and $A^{mr} \to A$. But

$$(A \to A^{mr})^{mr} \equiv \exists SY \forall v (v \ mr \ A \to \forall s \in Sv \# 0 \ Ysv \ mr \ A),$$

and S, Y defined by $Sv = 1$ and $Ysv = v$ mr-realize $A \to A^{mr}$;

$$(A^{mr} \to A)^{mr} \equiv \exists V_0 \forall SY (\forall s \in S \# 0 \ Ys \ mr \ A \to V_0 SY \ mr \ A)$$

This is a form of the absorption lemma 3.4.7: there is a term tuple $v_0[S, Y]$ such that

$$\vdash \forall SY (\forall s \in S \# 0 \ Ys \ mr \ A \to v_0[S, Y] \ mr \ A),$$

and $V_0 := \lambda SY.v_0[S, Y]$ mr-realizes $A^{mr} \to A$.

In analogy to the procedure in section 1.6, we characterize the modified realization translations of $CZF^{\omega-}$ by appropriate schemata of choice and of independence of premise.

3.4.11   Schemata of weak choice and of weak independence of premise

(1) An **axiom schema of weak choice** WAC is formulated in 3.3.15.

(2) The **∃free axiom schema of weak choice** is this schema WAC restricted to ∃free A:

$$WAC_{\exists free} \qquad \forall x \exists y A[x, y] \to \exists SY \forall x (Sx \# 0 \wedge \forall s \in Sx \ A[x, Ysx])$$

with ∃free A and non-empty tuples x, y of variables.

(3) The **∃free schema of weak independence of premise** is

$$WIP_{\exists free} \qquad (A \to \exists y B[y]) \to \exists XY (A \to X \# 0 \wedge \forall x \in X \ B[Yx])$$

with ∃free $A, B[y]$ and a non-empty tuple y of variables.

(4) The **schema of weak independence of Harrop premise** WIP_H is like (3), however, with A Harrop and $B[y]$ any formula.

(5) Abbreviation: $CZF^{\omega\top} :\equiv CZF^{\omega-} + WAC_{\exists free} + WIP_{\exists free}$

3.4.12   Lemma

1. $CZF^{\omega-}$ $\qquad$ $\vdash A^{mrG}$ $\qquad$ $\leftrightarrow A$ for $A \in L(T_\in)$

2. $CZF^{\omega-}$ $\qquad$ $\vdash (A \wedge B)^{mrG}$ $\quad$ $\leftrightarrow (A^{mrG} \wedge B^{mrG})$

3. $CZF^{\omega-}$ $\qquad$ $\vdash (A \to B)^{mrG}$ $\quad$ $\to (G \to A \to B) \wedge (A^{mrG} \to B^{mrG})$

4. $CZF^{\omega-}$ $\qquad$ $\vdash (\forall u A[u])^{mrG}$ $\quad$ $\to \forall u (A[u]^{mrG})$

5. $CZF^{\omega-} + WAC$ $\quad$ $\vdash \forall u (A[u]^{mrG})$ $\quad$ $\to (\forall u A[u])^{mrG}$

6. $CZF^{\omega-}$ $\qquad$ $\vdash (\exists u A[u])^{mrG}$ $\quad$ $\leftrightarrow \exists u (A[u]^{mrG})$

7. $CZF^{\omega\top}$ $\qquad$ $\vdash (A^{mr} \to B^{mr}) \to (A \to B)^{mr}$

8. $CZF^{\omega-} + WAC_{\exists free}$ $\vdash \forall u (A[u]^{mr})$ $\quad$ $\to (\forall u A[u])^{mr}$

Proof. As in the proof of 1.6.11, 1. to 4. are left to the reader.

5. and 8. By WAC, $\forall u \exists v \, v \, mrG \, A[u]$ implies

$\exists SV' \forall u \, Su\#0 \wedge \forall s \in Su \, V'su \, mrG \, A[u]$

and by absorption of bounds 3.4.7, there is a tuple V such that $\forall u \, Vu \, mrG \, A[u]$; hence 5.

In 8., $v \, mr \, A[u]$ is $\exists$free, so that in this case, $WAC_{\exists free}$ suffices.
6. is shown as in 3.3.4.

7. Since $v \, mr \, A$ and $y \, mr \, B$ are $\exists$free,

$\vdash (\exists v(v \, mr \, A) \to \exists y(y \, mr \, B)) \leftrightarrow \forall v(v \, mr \, A \to \exists y(y \, mr \, B))$

$\to \forall v \exists S, Y'(v \, mr \, A \to \forall s \in S\#0 \, Y's \, mr \, B)$ $\quad$ by $WIP_{\exists free}$

$\to \forall v \exists y_0(v \, mr \, A \to y_0 \, mr \, B)$ $\quad$ by absorption of bounds 3.4.7

$\to \exists S, Y_0 \forall v(v \, mr \, A \to \forall s \in Sv\#0 \, Y_0 sv \, mr \, B)$ $\quad$ by $WAC_{\exists free}$

$\to \exists Y \forall v(v \, mr \, A \to Yv \, mr \, B)$ $\quad$ by absorption of bounds.

That proves 7.

3.4.13   Proposition

In $CZF^{\omega-} + \{C^{mrG} \leftrightarrow C\}$, the schemata WAC and WIP_H are derivable.

Proof of WAC.

$(\forall x \exists y A[x,y])^{mrG} \equiv \exists SYV \forall x \forall s \in Sx\#0 \, Vsx \, mrG \, A[x, Ysx]$

$\qquad \equiv \exists SY(\forall x \forall s \in Sx\#0 \, A[x, Ysx])^{mrG}$,

and WAC follows from two axioms $C^{mrG} \leftrightarrow C$.

Proof of WIP_H. By 3.4.5 (1), $A^{mrG} \equiv \emptyset \, mrG \, A$, and $(A \to \exists y B[y])^{mrG}$ is
$(G \to A \to \exists y B[y]) \wedge \exists X, Y, Y'(\emptyset \, mrG \, A \to \forall x \in X\#0 \, Y'x \, mrG \, B[Yx])$
which implies

$\exists X, Y(A^{mrG} \to (\forall x \in X\#0 \, B[Yx])^{mrG})$

Thus, three axioms $C^{mrG} \leftrightarrow C$ yield WIP_H.

These results lead to a satisfactory characterization of modified realization mr in $CZF^{\omega-}$:

3.4.14 Characterization and extended realization theorem for mr

The theories

$$CZF^{\omega-} + \{C^{mr} \leftrightarrow C\}$$
$$CZF^{\omega-} + WAC_{\exists free} + WIP_{\exists free}$$
$$CZF^{\omega-} + WAC + WIP_H$$

are equivalent, and they are mr-maximal relative to $CZF^{\omega-}$.

Proof. $CZF^{\omega-} + \{C^{mr} \leftrightarrow C\}$ proves WAC and WIP_H by proposition 3.4.13. $CZF^{\omega-} + WAC + WIP_H$ contains $CZF^{\omega-} + WAC_{\exists free} + WIP_{\exists free}$, and this theory proves $C^{mr} \leftrightarrow C$, as is seen by subformula induction on C, with all induction steps being contained in lemma 3.4.12, 1. to 4. and 6. for $G \equiv \bot$, and 7., 8. mr-maximality is shown for the first theory in corollary 3.4.10.

Hybrid interpretations mrt and mq show closure of $CZF^{\omega-}$ under rules corresponding to the weak axiom schemata introduced above. We apply mrt-realization to prove a weak form of explicit definability in parameters (WED) in $CZF^{\omega-}$ and closure of $CZF^{\omega-}$ under rules corresponding to the axiom schemata WAC and WIP_H.

3.4.15 Theorem: Admissible rules

$CZF^{\omega-}$ is closed under weak explicit definability and under explicit forms of a rule of weak choice and a rule of weak independence of Harrop premise:

(WED) For any formula $B[y]$, if $\vdash \exists y\, B[y]$, then there are terms $X : o, Y$ such that

$$\vdash X \# 0 \wedge \forall x \in X\, B[Yx]$$

(Rule-WAC) If $\vdash \forall x^\sigma \exists y^\tau A[x,y]$, then there are terms S, Y such that

$$\vdash Sx \# 0 \wedge \forall s \in Sx\, A[x, Ysx]$$

(Rule-WIP_H) If $\vdash A \to \exists y B[y]$ and A is Harrop, then there are terms $X : o, Y$ such that

$$\vdash A \to X \# 0 \wedge \forall x \in X\, B[Yx]$$

Proof. *(WED)* is the special case $A \equiv \top$ of *(Rule-WIP_H)*. The premise of *(Rule-WAC)* implies $\exists y\, A[x,y]$. Therefore, by *(WED)*, there are terms $S'[x], Y'[x]$ such that $\vdash \forall s \in S'[x] \# 0\, A[x, Y'[x]s]$. So the terms $S := \lambda x. S'[x], Y := \lambda s x. Y'[x]s$ satisfy the conclusion of *(Rule-WAC)*. It remains to show *(Rule-WIP$_H$)*.

Let $\vdash A \rightarrow \exists y B[y]$ for a Harrop formula A. By *mrt*-realization 3.4.9, there are terms X, Y, V such that $CZF^{\omega-} \vdash X, Y, V \; mrt \; (A \rightarrow \exists y B[y])$, i.e.

$$CZF^{\omega-} \vdash (A \rightarrow \exists y B[y]) \wedge (\emptyset \; mrt \; A \rightarrow \forall x \in X \# 0 \; Vx \; mrt \; B[Yx])$$

The first conjunct is redundant, and since A is Harrop, $\vdash A \rightarrow \emptyset \; mrt \; A$ by lemma 3.4.5 (3) and

$$\forall x \in X \# 0 \; Vx \; mrt \; B[Yx] \rightarrow \forall x \in X \# 0 \; B[Yx]$$

by proposition 3.4.2. Therefore, as claimed,

$$CZF^{\omega-} \vdash A \rightarrow \forall x \in X \# 0 \; B[Yx]$$

(WED) is, at least apparently, weaker than (ED) which could be shown to hold for HA^ω in theorem 1.6.18. In the presence of (WED), however, (ED) is equivalent to its following special case:

$(ED)_0$ *For any term $X : o$, if $\vdash X \# 0$, then $\vdash t \in X$ for some term $t : o$.*

3.4.16   Lemma

If $CZF^{\omega-}$ satifies $(ED)_0$, then $CZF^{\omega-}$ also satisfies (ED).

Proof. Let $CZF^{\omega-} \vdash \exists y B[y]$. By (WED), there are terms $X : o, Y$ such that $CZF^{\omega-} \vdash X \# 0 \wedge \forall x \in X \; B[Yx]$. In particular, $CZF^{\omega-} \vdash X \# 0$, and by $(ED)_0$, there is a term $t : o$ such that $CZF^{\omega-} \vdash t \in X$. Together with $CZF^{\omega-} \vdash \forall x \in X \; B[Yx]$, this implies $CZF^{\omega-} \vdash B[Yt]$, and Yt is the desired term.

For the type o theory CZF^-, Rathjen [34] proves an existence property (EP) in the following form:

(EP) If $\vdash \exists x \, A[x]$ holds for a formula $A[x]$ having at most the free variable x, then there is a formula $C[x]$ with exactly x free such that

$$\vdash \exists ! \, x (C[x] \wedge A[x]).$$

He derives this existence property for CZF^- from a weak existence property (wEP) which is essentially a closed version of (WED), by an ordinal analysis of CZF^-. He conjectures that this ordinal analysis can be refined to yield a proof of (EP) *in parameters*, i.e. with further free variables besides x being allowed to occur in the fromulae $A[x]$ and $C[x]$ mentioned in (EP) above. If that is the case, then it may also be conjectured that this argument may be transferred from CZF^- to $CZF^{\omega-}$ with the result that $CZF^{\omega-}$ satisfies

(ED), contradicting conjectures in [13] and [14].
$CZF^{\omega-}$ does certainly not satisfy $(Rule\text{-}IP_H)$, because applied to the provable formula

$$x\#0 \to \exists y\ y \in x,$$

it would call for a definable choice functional $\mathsf{F}_C : o \to o$ satisfying

$$x\#0 \to \mathsf{F}_C x \in x$$

which is known not to exist in $CZF^{\omega-}$.

As in section 1.6, it remains to extend and characterize mrG-realization - and thereby the hybrids mrt and mq. In analogy to $HA^{\omega G}$, we introduce the family of extensions $CZF^{\omega G}$ of $CZF^{\omega-}$.

3.4.17 The theories $CZF^{\omega G}$

For a class Δ of formulae, as in 1.6.19, put $(G \to \Delta) := \{G \to D \mid D \in \Delta\}$ and define

$$CZF^{\omega G} :\equiv CZF^{\omega-} + (G \to WAC_{\exists free}) + (G \to WIP_{\exists free})$$

We have $CZF^{\omega\perp} \equiv CZF^{\omega-}$ and, as already mentioned under 3.4.11 (5), $CZF^{\omega\top} \equiv CZF^{\omega-} + WAC_{\exists free} + WIP_{\exists free}$. With this notation, mrG-realization 3.4.8 extends to:

3.4.18 Extended mrG-realization theorem

$$CZF^{\omega\top} \overset{mrG}{\leadsto} CZF^{\omega G}$$

Proof. In view of theorem 3.4.8, it remains to mrG-realize $WAC_{\exists free}$ and $WIP_{\exists free}$ in $CZF^{\omega G}$.

mrG-realization of $WAC_{\exists free}$. For $\exists$free A, let
$C :\equiv \forall x \exists y A[x,y]$ and $D :\equiv \exists S, Y \forall x \forall s \in Sx\#0\ A[x, Ysx]$. Then
$S, Y\ mrG\ C \equiv \forall x \forall s \in Sx\#0\ \emptyset\ mrG\ A[x, Ysx]$, and
$Z, S', Y'\ mrG\ D \equiv \forall z \in Z\#0 \forall x \forall s' \in S'zx\#0\ \emptyset\ mrG\ A[x, Y'zs'x]$. So,

$$CZF^{\omega G} \vdash S, Y\,mrG\ C \to 1, \lambda z.S, \lambda z.Y\ mrG\ D,$$

and since $G \to C \to D$ is an axiom $G \to WAC_{\exists free}$, $C \to D$ is by lemma 3.4.5 (7) mrG-realizable in $CZF^{\omega G}$.

mrG-realization of $WIP_{\exists free}$. For $\exists$free A, B, let
$C \equiv A \to \exists y B[y]$ and $D \equiv \exists XY\ D_0[X, Y] \equiv \exists XY (A \to \forall x \in X\#0\ B[Yx])$. Then

X, Y mrG $C \equiv (G \to C) \wedge E[X,Y]$ with
$E[X,Y] :\equiv (\emptyset\ mrG\ A \to \forall x \in X \#0\ \emptyset\ mrG\ B[Yx])$,
$\emptyset\ mrG\ D_0[X,Y] \equiv (G \to D_0[X,Y]) \wedge E[X,Y]$, and
S, X', Y' mrG $D \equiv \forall s \in S\#0((G \to D_0[X's, Y's]) \wedge \emptyset\ mrG\ D_0[X's, Y's])$.
$G \to C \to D$ is an axiom $G \to WIP_{\exists free}$. $D_0[X,Y]$, being $\exists$free, is by
lemma 3.4.5 (2) in $CZF^{\omega-}$ equivalent to $E[X,Y]$. So,
$$CZF^{\omega G} \vdash X, Y mrG\ C \to 1, \lambda s.X, \lambda s.Y\ mrG\ D,$$
and since $G \to C \to D$ is an axiom $G \to WIP_{\exists free}$, $C \to D$ is by lemma
3.4.5 (7) mrG-realizable in $CZF^{\omega G}$.

In the same way in which theorem 3.4.8 implies theorem 3.4.9, the extended
mrG-realization theorem implies:

3.4.19   Extended uniform realization theorem

$$CZF^{\omega-} + WAC_{\exists free} + WIP_{\exists free} \overset{mr}{\hookrightarrow} CZF^{\omega-}$$
$$CZF^{\omega-} + WAC_{\exists free} + WIP_{\exists free} \overset{mrt}{\hookrightarrow} CZF^{\omega-} + WAC_{\exists free} + WIP_{\exists free}$$
$$CZF^{\omega-} + WAC_{\exists free} + WIP_{\exists free} \overset{mq}{\hookrightarrow} CZF^{\omega-} + WAC_{\exists free} + WIP_{\exists free}.$$

Again, the same terms realize the same theorems under all three translations.

For mr, this result is already contained in theorem 3.4.14, due to the fact
that the schema $\{A \leftrightarrow A^{mr}\}$ is — via the absorption lemma — easily mr-
realized in $CZF^{\omega-}$. For the translations mrG in general, it is the extended
realization theorem 3.4.18 that leads to characterizations of mrG:

3.4.20   Proposition

$$CZF^{\omega\top} \vdash A^{mrG} \leftrightarrow A$$

Proof. By the characterization theorem for mr 3.4.14,

(1) $CZF^{\omega\top} \vdash A^{mr} \leftrightarrow A$

By extended mrG-realization 3.4.18, $A^{mr} \leftrightarrow A$ is therefore mrG-realizable
in $CZF^{\omega G}$ and a fortiori in $CZF^{\omega\top}$. This implies

(2) $CZF^{\omega\top} \vdash (A^{mr} \leftrightarrow A)^{mrG}$

so that by lemma 3.4.12

(3) $CZF^{\omega\top} \vdash (A^{mr})^{mrG} \leftrightarrow A^{mrG}$

But $A^{mr} \equiv \exists v\ v\ mr\ A$ with $\exists$free matrix $v\ mr\ A$. Therefore, by lemma 3.4.5,

(4) $CZF^{\omega-} \vdash (A^{mr})^{mrG} \leftrightarrow A^{mr}$

(1), (4), and (3) taken together yield the proposition.

3.4.21   Proposition

$$CZF^{\omega\top} \vdash A^{mq} \leftrightarrow A$$

Proof. By proposition 3.4.2, the implication $A \to A^{mq}$ is equivalent to $A^{mrt} \leftrightarrow A$, and its derivability is therefore equivalent to the case $G \equiv \top$ of the previous proposition. The implication $A^{mq} \to A$ follows by subformula induction. It suffices to show the case for implication:
$(A \to B)^{mq}$ implies $(A \wedge A^{mq}) \to B^{mq}$. But $\vdash A \to (A \wedge A^{mq})$ by the implication already shown, and $\vdash B^{mq} \to B$ by I.H. Hence,
$\vdash (A \to B)^{mq} \to (A \to B)$.

Putting these results together, we obtain:

3.4.22   Characterization theorem for mrG and mq

For any sentence G, the theories

$$CZF^{\omega-} + \{A^{mr} \leftrightarrow A\}$$
$$CZF^{\omega-} + \{A^{mrt} \leftrightarrow A\}$$
$$CZF^{\omega-} + \{A^{mq} \leftrightarrow A\}$$
$$CZF^{\omega-} + \{A^{mrG} \leftrightarrow A\}$$
$$CZF^{\omega-} + WAC_{\exists free} + WIP_{\exists free} \equiv CZF^{\omega\top}$$
$$CZF^{\omega-} + WAC + WIP_H$$

are equivalent.

Proof. By the characterization theorem 3.4.14 for mr, the theories in the first and in the last two lines are equivalent. Therefore and by propositions 3.4.20 and 3.4.21, $CZF^{\omega\top}$ proves at least as much as any theory in the list above. On the other hand, by proposition 3.4.13, for any sentence G, $CZF^{\omega-} + \{A^{mrG} \leftrightarrow A\}$ proves $WAC + WIP_H$. Finally, by proposition 3.4.2, already $CZF^{\omega-} + \{A \to A^{mq}\}$ proves $A^{mrt} \leftrightarrow A$ and therefore, by proposition 3.4.13 for $G \equiv \top$, again $WAC + WIP_H$.

3.5   Hybrids of the $\wedge$-interpretation of $CZF^{\omega-}$

Hybrid functional interpretations of $CZF^{\omega-}$ are developed here in analogy to the q- and t-hybrids of the $\wedge$-interpretation of HA^{ω} in section 1.7.

Transferring Stein's $\wedge q$-interpretation of HA^{ω} (cf. section 1.7) to $CZF^{\omega-}$, Schulte [35] develops a q-hybrid $\times q$ of his $\times$-interpretation of $CZF^{\omega-}$. We recast it as a $\wedge q$-interpretation.

Parallel to this, we transfer the t-hybrid inspired by Gaspar and Oliva [20] to a $\wedge t$-interpretation of $CZF^{\omega-}$ in close relation to the procedure in section 1.7.

3.5.1 *Recursive definition of hybrid functional translations* $\wedge q$ *and* $\wedge t$ *of* $CZF^{\omega-}$

(1) $\wedge q$ is defined like $\wedge$ in definition 3.2.4, with $\wedge$ replaced by $\wedge q$, and the following changes and additions:

$(\to)^{\wedge q}$ $(A \to B)^{\wedge q} \equiv \exists XWY \forall vz (A \wedge \forall x \in Xvz \, A_{\wedge q}[v, Wxvz] \to B_{\wedge q}[Yv, z])$
   in case the tuple w is not empty

$(\to)_0^{\wedge q}$ $(A \to B)^{\wedge q} \equiv \exists Y \forall vz (A \wedge A_{\wedge q}[v] \to B_{\wedge q}[Yv, z])$ for empty w

$(\exists)^{\wedge q}$ $(\exists u A[u])^{\wedge q} \equiv \exists XUV \forall w (X \# 0 \wedge \forall x \in X(A[Ux] \wedge A_{\wedge q}[Ux, Vx, w]))$

(2) $\wedge t$ is again defined like $\wedge$ in definition 3.2.4, with $\wedge$ replaced by $\wedge t$, and the following changes and additions:

$(\to)^{\wedge t}$ $(A \to B)^{\wedge t} \equiv \exists XWY \forall vz ((A \to B) \wedge (A \wedge \forall x \in Xvz \, A_{\wedge t}[v, Wxvz]$
   $\to B_{\wedge t}[Yv, z]))$
   in case the tuple w is not empty

$(\to)_0^{\wedge t}$ $(A \to B)^{\wedge t} \equiv \exists Y \forall vz ((A \to B) \wedge (A_{\wedge t}[v] \to B_{\wedge t}[Yv, z]))$ for empty w

$(\forall)^{\wedge t}$ $(\forall u A[u])^{\wedge t} \equiv \exists V \forall uw (\forall u A[u] \wedge A_{\wedge t}[u, Vu, w])$

$(\exists)^{\wedge t}$ $(\exists u A[u])^{\wedge t} \equiv \exists XUV \forall w (X \# 0 \wedge \forall x \in X \, A_{\wedge t}[Ux, Vx, w])$

These hybrid translations are related in the same way as in the arithmetic case in proposition 1.7.3:

3.5.2 *Proposition*

$$CZF^{\omega-} \vdash A^{\wedge t} \leftrightarrow A \wedge A^{\wedge q}, \; even$$

$$CZF^{\omega-} \vdash A_{\wedge t}[v, w] \leftrightarrow A \wedge A_{\wedge q}[v, w]$$

Theorems of $CZF^{\omega-}$ *thus have identical* $\wedge q$- *and* $\wedge t$-*interpreting term tuples.*

The *proof* by subformula induction is almost identical to the proof of proposition 1.7.3, after substitution of $\in$ for $<$. The only – merely formal – difference is in the induction step for the existential quantifier:

$(\exists)_{\wedge t}$ $(\exists u A[u])_{\wedge t} \equiv X \# 0 \wedge \forall x \in X \, A_{\wedge t}[Ux, Vx, w]$
is by I.H. equivalent to
$X \# 0 \wedge \forall x \in X(A[Ux] \wedge A_{\wedge q}[Ux, Vx, w]) \equiv (\exists u A[u])_{\wedge q}$
which is equivalent to $\exists u A[u] \wedge (\exists u A[u])_{\wedge q}$.
The other induction steps may be taken from the proof of proposition 1.7.3.

In full analogy to our procedure in section 1.7, we shall show that, given a theorem A of $CZF^{\omega-}$, there are such term tuples b which not only $\wedge q$- and $\wedge t$-, but also $\wedge$-interpret A:

3.5.3 Theorem: Uniform $\wedge$-, $\wedge q$-, and $\wedge t$- interpretation of $CZF^{\omega-}$

Let $A^I \equiv \exists v \forall w A_I[v, w]$ for $I = \wedge,\ \wedge q,\ \wedge t$. If $CZF^{\omega-} \vdash A$, then there is a tuple b of terms from $L(T_\in)$ (of the same type tuple as v) for which at the same time

$$T_\in \vdash A_\wedge[b, w], \quad CZF^{\omega-} \vdash A_{\wedge q}[b, w], \quad and \quad CZF^{\omega-} \vdash A_{\wedge t}[b, w]$$

Therefore

$$CZF^{\omega-} \overset{\wedge}{\hookrightarrow} T_\in, \quad CZF^{\omega-} \overset{\wedge q}{\hookrightarrow} CZF^{\omega-} \quad and \quad CZF^{\omega-} \overset{\wedge t}{\hookrightarrow} CZF^{\omega-}$$

In view of proposition 3.5.2, it suffices to prove this theorem for $\wedge$ and $\wedge q$, and as in section 1.7, we prove it again for a family of translations $\wedge G$ subsuming both these translations.

3.5.4 Recursive definition of translations $\wedge G$ of $L(CZF^{\omega-})$

Let G be an arbitrary sentence of $L(CZF^{\omega-})$. To any formula A of $L(CZF^{\omega-})$, $\wedge G$ assigns a formula $A^{\wedge G} \equiv \exists v \forall w A_{\wedge G}[v, w]$ of $L(CZF^{\omega-})$ as follows:

$(T_\in)^{\wedge G} \qquad A^{\wedge G} \equiv A$ for $A \in L(T_\in)$

Let $A^{\wedge G}$ be as above and $B^{\wedge G} \equiv \exists y \forall z B_{\wedge G}[y, z]$; then

$(\wedge)^{\wedge G} \quad (A \wedge B)^{\wedge G} \quad \equiv \exists v y \forall w z (A_{\wedge G} \wedge B_{\wedge G})$

$(\to)^{\wedge G} \quad (A \to B)^{\wedge G} \quad \equiv \exists XWY \forall vz(((G \to A) \wedge$
$\qquad\qquad\qquad\qquad\qquad \forall x \in Xvz A_{\wedge G}[v, Wxvz]) \to B_{\wedge G}[Yv, z])$
$\qquad\qquad\qquad\qquad$ in case the tuple w is not empty

$(\to)_0^{\wedge G} \quad (A \to B)^{\wedge G} \quad \equiv \exists Y \forall vz(((G \to A) \wedge A_{\wedge G}[v]) \to B_{\wedge G}[Yv, z])$
$\qquad\qquad\qquad\qquad$ for empty w

$(\forall \in)^{\wedge G} \quad (\forall x \in tB[x])^{\wedge G} \equiv \exists Y \forall z\ \forall x \in t\ B_{\wedge G}[x, Yx, z]$

$(\forall)^{\wedge G} \quad (\forall u A[u])^{\wedge G} \quad \equiv \exists V \forall uw A_{\wedge G}[u, Vu, w]$

$(\exists)^{\wedge G} \quad (\exists u A[u])^{\wedge G} \quad \equiv \exists XUV \forall w(X \# 0 \wedge \forall x \in X((G \to A[Ux])$
$\qquad\qquad\qquad\qquad\qquad\qquad \wedge A_{\wedge G}[Ux, Vx, w]))$

As before, we have $\wedge \bot \equiv \wedge$ and $\wedge \top \equiv \wedge q$. With the exception of the clause $(\exists)^{\wedge G}$, this definition corresponds exactly to definition 1.7.5 of $\wedge G$ in the arithmetic case, with $<$ replaced by $\in$. We transfer methods from there as far as possible. The translation of the existential quantifier may again be simplified in special cases:

3.5.5　Lemma

The addition of the following clauses to definition 3.5.4 does not extend the concept of $\wedge G$-interpretability:

(1) $(\exists^o)^{\wedge G}$　$(\exists x^o A[x])^{\wedge G} \equiv \exists X(X \# 0 \wedge \forall x \in X \, A[x])$　*for*　$A[x] \in L(T_\in)$

(2) *the clause* $(\exists)^{\wedge G}$ *above, applied to non-empty tuples u of variables.*

Proof. (1) For $A[x^o]$ in $L(T_\in)$, $(\exists x^o A[x])^{\wedge G}$ and $(\exists x^o A[x])^{\wedge}$ have identical prefixes and equivalent matrices: The same terms $\wedge$- and $\wedge G$-interpret $\exists x^o A[x]$. So the lemma follows from lemma 3.3.6.

(2) holds for $\wedge G$ for the same reason as for mq in section 3.4. A detailed proof (by induction on the length of u) is left to the reader. The crucial point in the induction step is contained in the proof of proposition 3.4.2.

Also the absorption lemma for $\wedge G$ has practically the same proof as the absorption lemma 3.3.7 for $\wedge$:

3.5.6　Lemma: Absorption of bounds for $\wedge G$.

Let $A^{\wedge G}$ be $\exists v \forall w A_{\wedge G}[v, w]$. Given terms S, V (not containing variables from w) there are terms v_0 (also not containing variables from w) such that

$$CZF^{\omega -} \vdash S \# 0 \wedge \forall s \in S \, A_{\wedge G}[Vs, w] \to A_{\wedge G}[v_0, w]$$

This lemma opens the way to the $\wedge G$-interpretation:

3.5.7　Theorem: Uniform $\wedge G$-interpretation of $CZF^{\omega -}$

If $CZF^{\omega -} \vdash A$, then there is a tuple b of terms with $FV(b) \subset FV(A)$ such that for all sentences G:

$$CZF^{\omega -} \vdash A_{\wedge G}[b, w] \, , \ \textit{hence}$$

$$CZF^{\omega -} \overset{\wedge G}{\hookrightarrow} CZF^{\omega -}$$

The *proof* by induction on deductions is an amendment of the proof of the $\wedge$-interpretation theorem 3.3.8 along the same lines as the proof of the uniform $\wedge G$-interpretation theorem 1.7.6 in arithmetic amends the proof of the $\wedge$-interpretation theorem 1.2.7.: The computation of the interpreting terms remains unchanged, only the additional conjuncts $G \to \ldots$ must be controlled.

For the axioms and rules concerning the negative fragment, i.e. for $A1$ to $A7$, $A\bot$, Eq, $Q1$, $Q3$, this works in exactly the same way as in the proof of theorem 1.7.6. Moreover, since $A^{\wedge G} \equiv A$ for $A \in L(T_{\in})$, the axioms and rules of $T_{\in}$, including axioms $A8, Q2b$, and rule $(T\text{-}EXT)$ are $\wedge G$-interpreted trivially. It remains to extend the $\wedge$-interpretation of $A9, Q2, Q4b, Q4$, (T IND), and $(StrongColl)$ to a $\wedge G$-interpretation in $CZF^{\omega -}$. We amend the correponding induction steps in the proof of theorem 3.3.8 or, in the case of (T IND), of theorem 3.2.7. All derivations in this proof are in $CZF^{\omega -}$.

$A9$. By I.H. and case distinction 3.1.6, there are terms Y such that for $i \in 2$

$$\vdash ((G \to A_i) \wedge A_i) \to B_{\wedge G}[Yi, z], \text{ hence also } \vdash A_i \to B_{\wedge G}[Yi, z].$$

By the argument in the proof of 3.3.8, there are terms y_0 such that

$$\vdash A_0 \vee A_1 \to B_{\wedge G}[y_0, z],$$

and these terms y_0 a fortiori $\wedge G$-interpret $A_0 \vee A_1 \to B$.

$Q2$. As in the proof of 3.3.8, $Svz = X = 1, Wsvz = z, Uvx = b, Yvx = v$ also $\wedge G$-interpret $Q2$, because

$$\vdash ((G \to A[b]) \wedge \forall s \in 1 \ A_{\wedge G}[b, v, z]) \to \forall x \in 1 \#0 \ A_{\wedge G}[b, v, z]$$

$Q4b$. This is analogous to $A9$. By I.H. and λ-abstraction, there are terms Y s.t. $\vdash ((G \to x \in t \wedge A[x]) \wedge x \in t \wedge A[x]) \to B_{\wedge G}[Yx, z]$, hence also

$$\vdash x \in t \wedge A[x] \to B_{\wedge G}[Yx, z]$$

By the argument in the proof of 3.3.8, there are terms y_0 such that

$$\vdash \exists x \in tA[x] \to B_{\wedge G}[y_0, z],$$

and these terms $y_0 \ \wedge G$-interpret $\exists x \in tA[x] \to B$ in $CZF^{\omega -}$.

$Q4$. $A[u] \to B \vdash \exists u \ A[u] \to B$ for u not free in B.
By I.H., there are terms X_0, W_0, Y_0 s.t.
(1) $\vdash ((G \to A[u]) \wedge \forall x_0 \in X_0 uvz \ A_{\wedge G}[u, v, W_0 x_0 uvz]) \to B_{\wedge G}[Y_0 uv, z]$.
Let $(\exists uA[u])^{\wedge G} :\equiv \exists Su'v'\forall w\forall s \in S\#0((G \to A[u's]) \wedge A_{\wedge G}[u's, v's, w])$.
By the same argument and for the same terms X, W, Y as in the proof of 3.3.8, it follows (without using the conjunct $G \to \exists uA[u]$ of the antecedent)
(2) $\vdash (G \to \exists uA[u]) \wedge \forall x \in X Su'v'z\forall s \in S\#0$
$\qquad ((G \to A[u's]) \wedge A_{\wedge G}[u's, v's, WxSu'v'z]) \to B_{\wedge G}[YSu'v', z]$
which is $(\exists uA[u] \to B)_{\wedge G}[X, W, Y; S, u', v', z]$.
This also completes the proof of lemma 3.5.5.

$(StrongColl) \quad \forall x \in a \exists y \ F[x, y] \to \exists b \ F'[a, b] \quad$ where

$$F'[a, b] :\equiv \forall x \in a \exists y \in b \ F[x, y] \wedge \forall y \in b \exists x \in a \ F[x, y]$$

Case 1. $F[x,y]$ is Δ_0. Applying lemma 3.5.5, $\wedge G$-interpretation in this case reduces to $\wedge$-interpretation as treated in the proof of theorem 3.3.8.
Case 2. $F[x,y]$ is not Δ_0, $F[x,y]^{\wedge G} \equiv \exists v \forall w F_{\wedge G}[x,y,v,w]$. Using the same notation and the same abbreviations as in the proof of 3.3.8, we arrive at

$$(StrongColl)^{\wedge G} \equiv \exists ZWbS_0Y_0v_0S_1X_1v_1\forall SY vw_0w_1$$
$$\forall i \in Z+\big((G \to \forall x \in a\exists y\, F[x,y]) \wedge \forall x \in a\forall s \in Sx\#0(G \to F[x,Yxs])$$
$$\wedge F_{\wedge G}[x,Yxs,vxs,Wi+]\big)$$
$$\to \Big(\forall x \in a\ \forall s \in S_0x-\#0\left(Y_0xs- \in b- \wedge (G \to F[x,Y_0xs-])\right.$$
$$\wedge F_{\wedge G}[x,Y_0xs-,v_0xs-,w_0])$$
$$\wedge \forall y \in b-\forall s \in S_1y-\#0\left(X_1ys- \in a \wedge (G \to F[X_1ys-,y])\right.$$
$$\wedge F_{\wedge G}[X_1ys-,y,v_1ys-,w_1])\Big)\Big)$$

Again, we put $Z+ := 2$, $Wi+ := w_i$ for $i \in 2$, $b- := \{Yxs|s \in Sx, x \in a\}$, $S_0x- := Sx, Y_0xs- := Yxs, v_0xs- := vxs$, and finally for $i = 1$:
$S_1y- := \{< x,s >|\ y = Yxs, s \in Sx, x \in a\}$, $X_1ys- := (s)_0$ and $v_1ys- := v(s)_0(s)_1$.
By the argument in the proof of theorem 3.3.8, the antecedent of $(StrongColl)_{\wedge G}$ implies the $\wedge$-part of both conjuncts in the consequent, but the same argument also holds for the G-part:

$$\vdash \forall x \in a\forall s \in Sx\#0(G \to F[x,Yxs]) \to$$
$$\forall x \in a\forall s \in S_0x-\#0\left(Y_0xs- \in b- \wedge (G \to F[x,Y_0xs-])\right)$$
$$\wedge \forall y \in b-\forall s \in S_1y-\#0\left(X_1ys- \in a \wedge (G \to F[X_1ys-,y])\right)$$

Due to the possibly complex formula $F[x,y]$ (and G), this formula is derived in $CZF^{\omega-}$. The derivation, however, does not use $(StrongColl)$.

For the $\wedge G$-interpretation of

(T IND) $\forall u \in a\, F[u] \to F[a] \vdash F[t]$

we go back to the corresponding part of the proof of theorem 3.2.7. By I.H., there are terms and term tuples X, W, Y_0 such that

$$(1) \quad \begin{aligned} CZF^{\omega-} \vdash\ &(G \to \forall u \in a\, F[u]) \wedge \forall x \in Xavz\forall u \in a\, F_{\wedge G}[u,vu,Wxavz] \\ &\to F_{\wedge G}[a,Y_0va,z] \end{aligned}$$

We define terms Y by simultaneous transfinite recursion 3.1.13

$$Ya = Y_0(Y \upharpoonright a)a\ ,$$

substitute $Y \upharpoonright a$ for v in (1), and obtain

$\vdash (G \to \forall u \in aF[u]) \wedge \forall x \in Xa(Y \upharpoonright a)z\,\forall u \in a\ F_{\wedge G}[u, Yu, Wxa(Y \upharpoonright a)z]$
$\to F_{\wedge G}[a, Ya, z]$

This is not exactly of the form of the hypothesis in proposition 3.1.18. However, in view of the hypothesis $\vdash \forall u \in a\ F[u] \to F[a]$, this can be rearranged to

$\vdash \forall u \in a((G \to F[u]) \wedge \forall x \in Xa(Y \upharpoonright a)z\, F_{\wedge G}[u, Yu, Wxa(Y \upharpoonright a)z])$
$\to (G \to F[a]) \wedge F_{\wedge G}[a, Ya, z]$

By the argument in the proof of theorem 3.2.7, this implies by an application of generalized transfinite induction 3.1.18 as a theorem of $CZF^{\omega-}$

$$\vdash (G \to F[t]) \wedge F_{\wedge G}[t, Yt, z]$$

and again $Yt \wedge G$-interprets $F[t]$.

This completes the amendation of the proof of the $\wedge$-interpretation theorems 3.3.8 and 3.2.7. Thereby, also the proofs of the uniform $\wedge G$-interpretation theorem 3.5.7 and, in particular, of the uniform interpretation theorem 3.5.3 are completed.

Some results generalize easily from the $\wedge$-translation to all $\wedge G$-translations, e.g. lemma 3.3.4:

3.5.8 Lemma

In $CZF^{\omega-}$ is provable:

1. $(\forall wA)^{\wedge G} \quad \leftrightarrow \forall wA \quad for \quad A \in L(T_\in)$
2. $(A \wedge B)^{\wedge G} \quad \leftrightarrow (A^{\wedge G} \wedge B^{\wedge G})$
3. $(A \to B)^{\wedge G} \quad \to ((G \to A) \wedge A^{\wedge G} \to B^{\wedge G})$
4. $(\forall x \in t\ A[x])^{\wedge G} \to \forall x \in t(A[x]^{\wedge G})$
5. $(\forall uA[u])^{\wedge G} \quad \to \forall u(A[u]^{\wedge G})$
6. $(\exists uA[u])^{\wedge G} \quad \leftrightarrow \exists u((G \to A[u]) \wedge A[u]^{\wedge G})$

The *proof* is almost identical with the one under lemma 3.3.4.

We now concentrate on the case $G \equiv \top$, i.e. on properties of the translation $\wedge q$ which, as in the arithmetic case, is more closely related to mr and mq than to $\wedge$. Several results transfer directly from section 1.7 to the present situation.

3.5.9 Proposition

For Harrop formulae A,

$$CZF^{\omega-} \vdash A \to A_{\wedge q}[v, w]$$

and therefore also

$$CZF^{\omega-} \vdash A \to A^{\wedge q}$$

The *proof by subformula induction* may be taken over from proposition 1.7.8.

From this proposition, we again obtain closure of $CZF^{\omega-}$ under the weak rules formulated in 3.4.15 in the previous section:

3.5.10   *Theorem*

Also the $\wedge q$-interpretation shows closure of $CZF^{\omega-}$ under (Rule-WIP_H) in its explicit form.

Proof. Let $CZF^{\omega-} \vdash A \to \exists y B[y]$ with A Harrop. By the $\wedge q$-interpretation theorem 3.5.3, there are term tuples X, W, S, Y, Y_1 such that $CZF^{\omega-}$ proves

$$A \wedge \forall x \in X vz\, A_{\wedge q}[v, Wxvz] \to \forall s \in Sv\#0\, B[Ysv] \wedge B_{\wedge q}[Ysv, Y_1 sv, z]$$

Since A is Harrop, $\vdash A \to \forall x \in X vz\, A_{\wedge q}[v, Wxvz]$ by the above proposition 3.5.9, and therefore

$$CZF^{\omega-} \vdash A \to \forall s \in S0\#0\, B[Ys0]$$

where the first and third 0 again stand for tuples of 0-functionals.

So, $S0$ and $\lambda s.Ys0$ are the terms satisfying the conclusion of (*Rule-WIP_H*). (*WED*) and (*Rule-WAC*) now follow as in theorem 3.4.15.

As in arithmetic, the schema $\{A \to A^{\wedge q}\}$ does not add anything to the theory, if restricted to Harrop formulae A, but extended to the full language of $CZF^{\omega-}$, it implies the schemata $\{A \to A^{mq}\}$ and $\{A \to A^{mrt}\}$:

3.5.11   *Proposition*

In $CZF^{\omega-} + \{A \to A^{\wedge q}\}$, the schemata WIP_H, WAC, and $\{A^{\wedge q} \to A\}$ are provable.

Proof. 1. Let $C \equiv A \to \exists y\, B[y]$ with A Harrop. Then by the argument in proposition 3.5.9,

$$C^{\wedge q} \to \exists SY(A \to \forall s \in S\#0\, B[Ys])$$

So, $C \to C^{\wedge q}$ implies WIP_H.

2. Let $C \equiv \forall x \exists y A[x, y]$. Then

$$C^{\wedge q} \equiv \exists SYV \forall xw(\forall s \in S\#0(A[x, Ysx] \wedge A_{\wedge q}[x, Ysx, Vsx, w]))$$

So, $C^{\wedge q}$ implies $\exists SY \forall x \forall s \in Sx \#0\, A[x, Ysx]$, and $C \to C^{\wedge q}$ is WAC.

3. The schema $\{A^{\wedge q} \to A\}$ is proved by subformula induction on A. The cases for $A \in L(T_\in)$, conjunction, bounded and unbounded universal quantification are immediate.

($\exists$) holds by 3.5.8.6 (for $G \equiv \top$) even without I.H.

($\to$) By the axiom $A \to A^{\wedge q}$ and the I.H. $B^{\wedge q} \to B$, 3.5.8.3 (for $G \equiv \top$) implies

$$(A \to B)^{\wedge q} \to (A \to B)$$

As an immediate consequence of this proposition, we have:

Corollary *In $CZF^{\omega-} + \{A \to A^{\wedge q}\}$, also the schemata $WIP_H^{\wedge q}$ and $WAC^{\wedge q}$ are provable.*

Concerning WIP_H, this result can again be strengthened:

3.5.12 *Proposition*

WIP_H *is $\wedge q$-interpretable in $CZF^{\omega-}$*

We give the *proof* in some detail, because
- it may be read as a refinement of the proof of the corresponding arithmetical proposition 1.7.11, and
- it also yields another (complicated) proof of closure of $CZF^{\omega-}$ under $(Rule\text{-}WIP_H)$ (cf. theorem 3.5.10).

Let A be Harrop,

$$C \equiv A \to \exists y B[y], \quad D_0[X, Y] \equiv A \to \forall x \in X \#0\, B[Yx], \quad D \equiv \exists XY\, D_0[X, Y]$$

Then $C \to D$ is a formula WIP_H, and $(C \to D)^{\wedge q} \equiv$

$$\exists X_0 W_0 Y_0 \forall v_0 z_0 (C \wedge \forall x \in X_0 v_0 z_0\, C_{\wedge q}[v_0, W_0 x v_0 z_0] \to D_{\wedge q}[Y_0 v_0, z_0])$$

Let $B[y]^{\wedge q} \equiv \exists y_1 \forall z B_{\wedge q}[y, y_1, z]$. From here onwards, we prime the tuples of bound variables that will constitute the tuples X_0, W_0, Y_0, and leave unprimed the tuples constituting v_0, z_0. Then

$(\exists y B[y])^{\wedge q} \equiv \exists S Y Y_1 \forall z' \forall s \in S \#0 (B[Ys] \wedge B_{\wedge q}[Ys, Y_1 s, z'])$ and

$(\forall x \in X \#0\, B[Yx])^{\wedge q} \equiv \exists Y_1' \forall z \forall x \in X \#0\, B_{\wedge q}[Yx, Y_1's, z]$

Since A is Harrop, $\vdash A \to \forall x \in X_1 v' z' A_{\wedge q}[v', W_1 x v' z']$ by proposition 3.5.9, and prefix and matrix of $C^{\wedge q}$ may be simplified:

$$\vdash C^{\wedge q} \leftrightarrow \exists S Y Y_1 \forall v' z' (A \to \forall s \in Sv' \#0 (B[Yv's] \wedge B_{\wedge q}[Yv's, Y_1 v's, z']))$$

For the same reason, $D_0[X, Y]^{\wedge q}$ is simplified in the same way so that

$$\vdash D_0[X, Y]^{\wedge q} \leftrightarrow \exists Y_1' \forall v z (A \to \forall x \in X \#0\, B_{\wedge q}[Yx, Y_1'vx, z])$$

which yields

$$\vdash D^{\wedge q} \leftrightarrow \exists S' X' Y' Y'_1 \forall vz \forall s \in S' \#0(D_0[X's, Y's] \wedge D_{0 \wedge q}[X's, Y's, Y'_1 s; v, z])$$

$$\leftrightarrow \exists S' X' Y' Y'_1 \forall vz \forall s \in S\#0\big((A \rightarrow \forall x \in X's\#0\, B[Y'sx])$$

$$\wedge\, (A \rightarrow \forall x \in X's\#0\, B_{\wedge q}[Y'sx, Y'_1 svx, z])\big)$$

The tuples v_0, z_0 of universally bound variables in the prefix of $(C \rightarrow D)^{\wedge q}$ are

$$v_0 \equiv X_1, W_1, S, Y, Y_1 \text{ and } z_0 \equiv v, z$$

with $(C \rightarrow D)_{\wedge q}$ independent of X_1, W_1. We put

$$X_0 v_0 z_0 = 1 \text{ and } W_0 x v_0 z_0 = 0, z$$

where 0 is a tuple of 0-functionals of the same type tuple as v. It remains to find a term tuple $Y_0 \equiv S', X', Y', X'_1, W'_1, Y'_1$ such that

(1) $$\vdash C_{\wedge q}[v_0; 0, z] \rightarrow D_{\wedge q}[Y_0 v_0; v, z]$$

with

$$\vdash C_{\wedge q}[v_0; 0, z] \leftrightarrow (A \rightarrow \forall s \in S0\#0(B[Y0s] \wedge B_{\wedge q}[Y0s, Y_1 0s, z]))$$

and

$$\vdash D_{\wedge q}[Y_0 v_0; v, z] \leftrightarrow$$

$$\forall s \in S'\#0(A \rightarrow \forall x \in X's\#0(B[Y'sx] \wedge B_{\wedge q}[Y'sx, Y'_1 svx, z]))$$

For $S' = 1, X's = S0, Y'sx = Y0x, Y'_1 svx = Y_1 0x$ and arbitrary X'_1, W'_1, (1) becomes a logical identity.

Two points concerning the $\wedge q$-interpretation of WIP_H are perhaps worth mentioning:

(i) The Harrop property of A eliminates occurrences of v from the antecedent of $D_{\wedge q}[Y_0 v_0; v, z]$, allowing us to interpret $W_0 x v_0 z_0$ by $0, z$, thus eliminating v from $C_{\wedge q}[v_0; 0, z]$. This suggests that formulations of WIP with non-Harrop antecedents are not interpretable. That pertains already to the arithmetic case discussed under 1.7.11.

(ii) The bound $\forall s \in S0\#0$ in $C_{\wedge q}[v_0; 0, z]$ corresponds to the bound $\forall x \in X's\#0$ which essentially occurs already in D. This suggests that the stronger schema IP_H is not interpretable in $CZF^{\omega -}$.

We also conjecture that the full schema WAC is not $\wedge q$-interpretable and $(WAC)^{\wedge q}$ is underivable in $CZF^{\omega -} + WIP_H + WAC$. Restriction of these schemata to *dual Harrop* formulae, as in definition 1.7.12, will solve part of the characterization problem:

3.5.13 Dual Harrop or dH formulae, and related WAC- and WIP-schemata

Dual Harrop or **dH** formulae are formulae of $L(CZF^{\omega-})$ without strictly negative occurrences of (unbounded) existential quantifiers. They are recursively defined as under 1.7.12, with $L(T_\wedge)$ replaced by $L(T_\in)$. Moreover, we introduce the schemata

WAC_H $\forall x \exists y A[x,y] \to \exists SY \forall x \forall s \in Sx \#0\, A[x, Ysx]$ with A Harrop;

WAC_{dH} $\forall x \exists y A[x,y] \to \exists SY \forall x \forall s \in Sx \#0\, A[x, Ysx]$ with A dH;

WIP_{HdH} $(A \to \exists y B[y]) \to \exists XY (A \to \forall x \in X \#0\, B[Yx])$ with A Harrop, B dual Harrop.

Axioms WAC_{dH} as well as WIP_{HdH} are both implications between two dual Harrop formulae. By restricting the proof of proposition 3.5.11 to dual Harrop formulae, we obtain:

3.5.14 Corollary to proposition 3.5.11

In $CZF^{\omega-} + \{A \to A^{\wedge q} \mid A\ dual\ Harrop\}$, the schemata WIP_{HdH} and WAC_{dH} are derivable.

3.5.15 Lemma

If A is dual Harrop,

$$CZF^{\omega-} \vdash A^{\wedge q} \to A$$

The *proof* by subformula induction proceeds by 1,2,3,4,5 of lemma 3.5.8. In the case of implications $A \to B$, A is Harrop and B is dual Harrop, so that $CZF^{\omega-}$ proves $A \to A^{\wedge q}$ by proposition 3.5.9 and $B^{\wedge q} \to B$ by I.H. This, together with lemma 3.5.8.6, implies

$$CZF^{\omega-} \vdash (A \to B)^{\wedge q} \to (A \to B)$$

With respect to dual Harrop formulae, $\wedge q$ behaves like *mrt* with respect to the full language:

3.5.16 Lemma

1. $CZF^{\omega-} + WAC \vdash \forall u(A[u]^{\wedge q}) \to (\forall u A[u])^{\wedge q}$ *for all A*
2. $CZF^{\omega-} + WIP_H \vdash (A \to B^{\wedge q}) \to (A \to B)^{\wedge q}$ *if A is Harrop*

Proof. 1. $\forall u(A[u]^{\wedge q}) \equiv \forall u \exists v \forall w A_{\wedge q}[u, v, w]$ implies by WAC
$$\exists XV\forall u \ (Xu\#0 \wedge \forall x \in Xu \,\forall w \ A_{\wedge q}[u, Vxu, w]),$$
equivalently $\exists XV\forall uw \ (Xu\#0 \wedge \forall x \in Xu \ A_{\wedge q}[u, Vxu, w])$
which by absorption of bounds 3.5.6 implies
$$\exists v_0 \forall uw \ A_{\wedge q}[u, v_0 u, w] \equiv (\forall u A[u])^{\wedge q}$$
2. $(A \to B^{\wedge q}) \equiv (A \to \exists y \forall z B_{\wedge q}[y, z])$ implies by WIP_H, since A is Harrop,
$$\exists XY(A \to X\#0 \wedge \forall x \in X \,\forall z \ B_{\wedge q}[Yx, z])$$
equivalently $\exists XY(A \to \forall z \ X\#0 \wedge \forall x \in X \ B_{\wedge q}[Yx, z])$ which by absorption
of bounds 3.5.6 implies
$$\exists y_0(A \to \forall z \ B_{\wedge q}[y_0, z])$$
which logically implies $(A \to B)^{\wedge q}$.

3.5.17 Partial characterization theorem for $\wedge q$

The theories
$$CZF^{\omega -} + \{A \leftrightarrow A^{\wedge q} \mid A \text{ dual Harrop }\}$$
$$CZF^{\omega -} + WAC_{dH} + WIP_{HdH}$$
$$CZF^{\omega -} + WAC_{\exists free} + WIP_{\exists free}$$
$$CZF^{\omega -} + \{A \leftrightarrow A^{mr}\}$$
$$CZF^{\omega -} + WAC + WIP_H$$
are equivalent.

Proof. By corollary 3.5.14, $CZF^{\omega -} + \{A^{\wedge q} \leftrightarrow A \mid A \text{ dual Harrop }\}$ proves
the schemata $WAC_{dH} + WIP_{HdH}$ which contain the schemata $WAC_{\exists free} +$
$WIP_{\exists free}$, because $\exists$free formulae are Harrop as well as dual Harrop. These
last schemata are, by the characterization theorem for mr 3.4.14, equivalent
to $\{A \leftrightarrow A^{mr}\}$ and to $WAC + WIP_H$ which in turn contain $WAC_{dH} +$
WIP_{HdH}. Because of lemma 3.5.15, it remains to show for dual Harrop
formulae A
$$CZF^{\omega -} + WAC + WIP_H \vdash A \to A^{\wedge q}$$
This is done by subformula induction. Because of lemma 3.5.8, it suffices to
show the cases for ($\forall$) and ($\to$).
($\forall$) follows from I.H. by lemma 3.5.16.1.
($\to$) : Given A Harrop and B dual Harrop, $\vdash B \to B^{\wedge q}$ by I.H., thus
$\vdash (A \to B) \to (A \to B^{\wedge q})$, and
$$CZF^{\omega -} + WAC + WIP_H \vdash (A \to B) \to (A \to B)^{\wedge q}$$
by lemma 3.5.16.2 - this is the point where $A \to B$ must be dual Harrop.
The case $\forall \in$ is an obvious combination of the last two cases.

It remains to close the gap between this partial and a full characterization
theorem.

3.5.18 Lemma

Let Γ be a class of formulae from which the schema WAC is provable. Then

$$CZF^{\omega^-} + \Gamma \vdash A \text{ implies } CZF^{\omega^-} + \Gamma + \Gamma^{\wedge q} \vdash A^{\wedge q}$$

Proof. By the deduction theorem, applied to the premise, there are $C_1, ..., C_n \in \Gamma$ such that

$$CZF^{\omega^-} \vdash \forall C_1 \wedge ... \wedge \forall C_n \to A$$

By the $\wedge q$-interpretation theorem 3.5.3,

$$CZF^{\omega^-} \vdash (\forall C_1 \wedge ... \wedge \forall C_n \to A)^{\wedge q}$$

This implies by lemma 3.5.8,3 and 2

$$CZF^{\omega^-} \vdash \forall C_1 \wedge ... \wedge \forall C_n \wedge (\forall C_1)^{\wedge q} \wedge ... \wedge (\forall C_n)^{\wedge q} \to A^{\wedge q}$$

and therefore

$$CZF^{\omega^-} + \Gamma \vdash (\forall C_1)^{\wedge q} \wedge ... \wedge (\forall C_n)^{\wedge q} \to A^{\wedge q}.$$

Since $CZF^{\omega^-} + \Gamma$ proves WAC, we may by lemma 3.5.16.1 replace $(\forall C_i)^{\wedge q}$ by $\forall (C_i^{\wedge q})$ for $i = 1, ..., n$ and obtain

$$CZF^{\omega^-} + \Gamma + \Gamma^{\wedge q} \vdash A^{\wedge q}$$

3.5.19 Axiomatization theorem for $\{A \to A^{\wedge q}\}$

For all A,

$$CZF^{\omega^-} + WIP_{\exists free} + WAC_{\exists free} + (WAC_{\exists free})^{\wedge q} \vdash A \to A^{\wedge q}$$

Proof. By theorem 3.4.14,

(1) $CZF^{\omega^-} + WIP_{\exists free} + WAC_{\exists free} \vdash A^{mr} \to A$

and the same theory proves the schema WAC. Therefore, lemma 3.5.16 is applicable, and since $CZF^{\omega^-} \vdash WIP_{\exists free}^{\wedge q}$ by proposition 3.5.12,

(2) $CZF^{\omega^-} + WIP_{\exists free} + WAC_{\exists free} + (WAC_{\exists free})^{\wedge q} \vdash (A^{mr} \to A)^{\wedge q}$

By lemma 3.5.8.3,

(3) $CZF^{\omega^-} \vdash (A^{mr} \to A)^{\wedge q} \to (A^{mr} \wedge (A^{mr})^{\wedge q} \to A^{\wedge q})$

But $v \; mr \; A$ is $\exists$free and therefore Harrop, so that by proposition 3.5.9 $CZF^{\omega^-} \vdash v \; mr \; A \to (v \; mr \; A)^{\wedge q}$ and by 3.5.8.6 also

$$CZF^{\omega^-} \vdash A^{mr} \to (A^{mr})^{\wedge q}$$

Therefore, $(A^{mr})^{\wedge q}$ may be cancelled in (3) so that

$$CZF^{\omega-} \vdash (A^{mr} \to A)^{\wedge q} \to (A^{mr} \to A^{\wedge q})$$

and by (2)

(4) $CZF^{\omega-} + WIP_{\exists free} + WAC_{\exists free} + (WAC_{\exists free})^{\wedge q} \vdash A^{mr} \to A^{\wedge q}$

As this theory, by theorem 3.4.14, also proves $A \to A^{mr}$, the proof is completed.

In analogy to Stein's principle of independent choice IC_H, we formulate a weak choice principle:

3.5.20   The weak principle of independent choice in constructive set theory

WIC_H $\forall SY \exists ZX (\forall z \in Z \forall s \in S(Xz)\#0\ A[Xz, Y(Xz)s]$
$\qquad\qquad \to \exists S_1 Y_1 (\forall x \exists y A[x,y] \to \forall x \forall s \in S_1 x \#0\ A[x, Y_1 sx]))$
for all Harrop formulae A.

Here is an attempt to read this formula semi-formally: *For all pairs* $Y : \sigma \to o \to \tau$, $S : \sigma \to o$, *there is a "critical" set* $\{Xz \mid z \in Z\}$ *such that: If* Y, S *is a choice relation for* $A[x,y]$ *on that critical set, then a pair* Y_1, S_1 *can be found which is in fact a choice relation for all of* $A[x,y]$ *(on the universe of type* σ*), provided* $\forall x \exists y A[x,y]$.
Classically, this is an obvious fact, even without the proviso $\forall x \exists y A[x,y]$. Constructively, this principle characterizes $(WAC_H)^{\wedge q}$:

3.5.21   Proposition

On the basis of $CZF^{\omega-} + WIP_H + WAC$, *the schema* $(WAC_H)^{\wedge q}$ *is equivalent to the above weak principle of independent choice* WIC_H *for Harrop formulae.*

Proof. Let A be Harrop, $C \equiv \forall x \exists y A[x,y]$ and
$\qquad\qquad D \equiv \exists S_1 Y_1 \forall x \forall s \in S_1 x \#0\ A[x, Y_1 sx]$.
We have to compute $(C \to D)^{\wedge q}$, up to equivalence.

$$C^{\wedge q} \equiv \exists SYV \forall x w \forall s \in Sx \#0 (A[x, Yxs] \wedge A_{\wedge q}[x, Yxs, Vxs, w])$$

Because A is Harrop, by proposition 3.5.9,
$\vdash A[x, Yxs] \to A_{\wedge q}[x, Yxs, Vxs, w]$, and

(1) $C^{\wedge q} \leftrightarrow \exists SY \forall x \forall s \in Sx \#0\ A[x, Yxs]$

$$D^{\wedge q} \equiv \exists S_2 S_1 Y_1 V_1 \forall x_1 w \forall s_2 \in S_2 \#0 (\forall x \forall s \in S_1 s_2 x \#0\ A[x, Y s_2 sx]$$

$$\land \forall s \in S_1 s_2 x_1 \#0 \, A[x_1, Y s_2 s x_1, V_1 s_2 x_1 s, w])$$

Again, by proposition 3.5.9, the first conjunct implies the second conjunct, whence

(2) $D^{\wedge q} \leftrightarrow \exists S_2 S_1 Y_1 \forall s_2 \in S_2 \#0 \, \forall x \forall s \in S_1 s_2 x \#0 \, A[x, Y s_2 s x]$

In this formula, the order of quantifiers $\forall s_2 \in S_2 \#0 \, \forall x$ may be interchanged so that by contraction of bounds 3.1.12 , (2) is further simplified to

(3) $D^{\wedge q} \leftrightarrow \exists S_1 Y_1 (\forall x \forall s \in S_1 x \#0 \, A[x, Y s x])$

Here, in contrast to $C^{\wedge q}$, the universal quantifier $\forall x$ belongs to the matrix $D_{\wedge q}$ of $D^{\wedge q}$. Thus, $(W A C_H)^{\wedge q} \equiv (C \rightarrow D)^{\wedge q}$ is equivalent to

(4) $\exists Z X S_1 Y_1 \forall S Y (\forall x \exists y A[x, y] \land \forall z \in Z S Y \forall s \in S(X z S Y) \#0$
$\qquad A[X z S Y, Y(X z S Y) s] \rightarrow \forall x \forall s \in S_1 S Y x \#0 \, A[x, Y_1 S Y s x])$

which implies

(5) $\forall S Y \exists Z X S_1 Y_1 (\forall x \exists y A[x, y] \land \forall z \in Z \forall s \in S(X z) \#0 \, A[X z, Y(X z) s]$
$\qquad \rightarrow \forall x \forall s \in S_1 x \#0 \, A[x, Y_1 s x])$

which in turn implies $W I C_H$.

In the opposite direction, $W I C_H$ implies by an application of $W I P_H$

$$\forall S Y \exists Z X S_2 S_1 Y_1 (\forall z \in Z \forall s \in S(X z) \#0 \, A[X z, Y(X z) s]$$
$$\rightarrow \forall s_2 \in S_2 \#0 (\forall x \exists y A[x, y] \rightarrow \forall x \forall s \in S_1 x \#0 \, A[x, Y_1 s x]))$$

which, after interchanging $\forall s_2 \in S_2 \#0$ and $\forall x \exists y A[x, y] \rightarrow \forall x$ in the consequent, by contraction of bounds 3.1.12 implies (5). WAC applied to (5) yields

$$\exists S_0 Z X S_1 Y_1 \forall S Y \forall s_0 \in S_0 S Y \#0$$
$$(\forall x \exists y A[x, y] \land \forall z \in Z s_0 S Y \forall s \in S(X s_0 z S Y) \#0 A[X s_0 z S Y, Y(X s_0 z S Y) s]$$
$$\rightarrow \forall x \forall s \in S_1 s_0 S Y x \#0 \, A[x, Y_1 s_0 S Y s x])$$

After distribution of $\forall s_0 \in S_0 S Y \#0$ over the matrix of this formula, we have

$$\exists S_0 Z X S_1 Y_1 \forall S Y \left(\forall x \exists y A[x, y] \right.$$
$$\land \forall s_0 \in S_0 S Y \forall z \in Z s_0 S Y \forall s \in S(X s_0 z S Y) \#0 A[X s_0 z S Y, Y(X s_0 z S Y) s]$$
$$\rightarrow \forall x \forall s_0 \in S_0 S Y \#0 \forall s \in S_1 s_0 S Y x \#0 \, A[x, Y_1 s_0 S Y s x])$$

which by contraction of bounds becomes (4) or equivalently $(W A C_H)^{\wedge q}$. These last two steps are, in fact, an application of the absorption lemma to the matrix of (4), if we generously identify (4) with $(W A C_H)^{\wedge q}$.

These results yield an axiomatization of the schema $\{A \leftrightarrow A^{\wedge q}\}$ as well as of $\{A \rightarrow A^{\wedge q}\}$ by axiom schemata including $W I C_H$ which do not mention any functional translation:

3.5.22   Characterization theorem for $\wedge q$

The following theories are equivalent:

$$CZF^{\omega-} + \{A \leftrightarrow A^{\wedge q}\}$$
$$CZF^{\omega-} + \{A \rightarrow A^{\wedge q}\}$$
$$CZF^{\omega-} + WIP_{\exists free} + WAC_{\exists free} + (WAC_{\exists free})^{\wedge q}$$
$$CZF^{\omega-} + WIP_H + WAC + (WAC)^{\wedge q}$$
$$CZF^{\omega-} + WIP_H + WAC + (WAC_H)^{\wedge q}$$
$$CZF^{\omega-} + WIP_H + WAC + WIC_H$$

Proof. By proposition 3.5.11, the first two theories are equivalent and prove the axiom schemata of the next three theories. By the axiomatization theorem 3.5.19, each of these three theories proves the schema $\{A \rightarrow A^{\wedge q}\}$. The last two theories are equivalent by proposition 3.5.21.

As in section 1.7, we return to the comparison of $\wedge t$ and $\wedge q$. Also the characterization theorem 3.5.22 transfers from $\wedge q$ to $\wedge t$:

3.5.23   Characterization theorem for $\wedge t$

On the basis of $CZF^{\omega-}$, the schemata $\{A \leftrightarrow A^{\wedge t}\}$ and $\{A \leftrightarrow A^{\wedge q}\}$ are equivalent: The functional interpretations $\wedge t$ and $\wedge q$ are characterized by the same schemata listed in the characterization theorem for $\wedge q$ above.

The *proof* copies the argument of the proof of the corresponding characterization theorem 1.7.21, this time, of course, arguing in $CZF^{\omega-}$. Since $A^{\wedge t} \rightarrow A$ by proposition 3.5.2, $A \leftrightarrow A^{\wedge t}$ is equivalent to $A \rightarrow A^{\wedge t}$, hence again by proposition 3.5.2 to $A \rightarrow A \wedge A^{\wedge q}$. i.e. to $A \rightarrow A^{\wedge q}$. However, by proposition 3.5.11, the schema $\{A \rightarrow A^{\wedge q}\}$ is equivalent to the schema $\{A \leftrightarrow A^{\wedge q}\}$. So the characterization theorem also for $\wedge t$ is proved.

3.6   Type-extensionality and majorizability of constructive set functionals

In analogy to Howard's argument presented in section 1.10, Burr [6] shows that the axiom of type extensionality $(t - ext)$ is not $\times$-interpretable by constructive set functionals. His argument transfers to the $\wedge$-interpretation. We start with a brief discussion of the schema of *subset collection* and the *fullness axiom* in the first order extension CZF of CZF^-, following Aczel [1], in order to motivate the introduction of a fullness functional F extending $T_\in$ and $CZF^{\omega-}$.

3.6.1 The schema of subset collection and the theory CZF

(*Subset Collection*) $\exists c\, \forall u(\forall x \in a\, \exists y \in b\, F[x,y] \to \exists d \in c\, F'[a,d])$

where u may occur free in F and $F'[a,d]$ is defined as in 3.3.1 (3):

$$F'[a,d] \quad\equiv\quad \forall x \in a\, \exists y \in d\, F[x,y] \wedge \forall y \in d\, \exists x \in a\, F[x,y]$$

Aczel then defines:

$$CZF \quad:\equiv\quad CZF^- + (\textit{Subset Collection})$$

For the formulation of the *fullness axiom*, some elementary terminology is needed.

3.6.2 Set theoretic terminology

f is a **function** from a to b $\Leftrightarrow$ $f : a \to b$

$\Leftrightarrow$ $f \subset a \times b \wedge \forall x \in a\, \exists! y \in b\, \langle x,y \rangle \in f$

Here, $dom\, f := \{(z)_0 \mid z \in f\} = a$ and $im\, f := \{(z)_1 \mid z \in f\} \subset b$

r is an a-**total relation** between a and b $\Leftrightarrow$ $r : a - < b$

$\Leftrightarrow$ $r \subset a \times b \wedge \forall x \in a\, \exists y \in b\, \langle x,y \rangle \in r$

r is a **total relation** between a and b $\Leftrightarrow$ $r : a > - < b$

$\Leftrightarrow$ $r : a - < b \wedge \forall y \in b\, \exists x \in a\, \langle x,y \rangle \in r$

c is an a-**full set of subsets of** b

$\Leftrightarrow$ $\forall d \in c\, d \subset b \wedge \forall r(r : a - < b \to \exists d \in c\, r : a > - < d)$

3.6.3 The fullness axiom

(*Fullness*) $\forall ab\, \exists c(c$ is an a-full set of subsets of $b):$

$\forall ab\, \exists c\, \forall r(r : a - < b \to \exists d \in c\, r : a > - < d)$

Remark. A set c existing by this last formulation may not in itself be a set of subsets of b; by Δ_0-separation, however, it gives rise to a subset $c' = \{d \in c \mid d \subset b\}$ which then is an a-full set of subsets of b.

3.6.4 Proposition: Equivalence of subset collection and fullness

On the basis of CZF^-, the axiom schema of subset collection is equivalent to the fullness axiom.

Proof from left to right. Consider the instance $F[x,y] \equiv u \subset a \times b \wedge \langle x,y \rangle \in u$ of (*Subset Collection*). Then

$$\forall x \in a\, \exists y \in b\, F[x,y] \leftrightarrow u : a - < b \quad \text{and} \quad F'[a,c] \leftrightarrow u : a > - < c$$

Therefore, (*Subset Collection*) implies

$$\exists c \, \forall u (u : a - < b \to \exists d \in c \; u : a > - < d)$$

and the universal closure of this formula is the fullness axiom.

Proof from right to left. Let c be an a-full set of subsets of b, and assume

(1) $\forall x \in a \, \exists y \in b \, F[x, y]$. Define

(2) $G[x, z] :\equiv \exists y \in b \, (F[x, y] \wedge \langle x, y \rangle = z)$

Then, by (1), $\forall x \in a \, \exists z \, G[x, z]$ which implies by (*Strong Collection*)

(3) $\exists r (\forall x \in a \, \exists z \in r \, G[x, z] \wedge \forall z \in r \, \exists x \in a \, G[x, z])$.

For any such r, the second conjunct of (3) implies $r \subset a \times b$ so that by (1) and the first conjunct

(4) $r : a - < b$ and

(5) $\forall x y (\langle x, y \rangle \in r \to F[x, y])$

For the set c existing by (*Fullness*), (4) implies:

(6) $\exists d \in c \; r : a > - < d$

which by (5) implies $F'[a, d]$, as was to be shown.

3.6.5   The exponentiation axiom

For all a, b, there exists the set ${}^a b$ of all functions from a to b:

$$\forall a, b \, \exists c \, \forall f \, (f \in c \leftrightarrow f : a \to b)$$

This axiom is also a theorem of CZF:

3.6.6   *Proposition: Fullness implies exponentiation*

On the basis of CZF^-, the fullness axiom implies the exponentiation axiom.

Proof. Let c be an a-full set of subsets of $a \times b$. If $f : a \to b$, define f' by

$$f'(x) = \langle x, f(x) \rangle$$

Then $f' : a \to a \times b$, hence also $f' : a - < a \times b$.

Therefore, by (*Fullness*), $\exists d \in c \, f' : a > - < d$

Since f' is a function on a, $d = \{ f'(x) \mid x \in a \} = f$ and $f \in c$. So the set c contains all functions from a to b, and by Δ_0-separation, we obtain as a set

$${}^a b := \{ f \in c \mid f : a \to b \}$$

Rather than the schema of subset collection, the fullness axiom lends itself to a functional interpretation, simply by the introduction of a *fullness functional* F which, for given a, b, throws out an a-full set $\mathsf{F}ab$ of subsets of b:

3.6.7 The theories $T_\in^+$ and CZF^ω, the fullness functional F and the exponentiation functional exp

The theory $T_\in^+$ is the extension of the theory $T_\in$ obtained by adding the **fullness functional** F $: o \to o \to o$ satisfying the **fullness axiom**

$$(F) \qquad r : a - < b \to \exists d \in \mathsf{F}ab \ r : a > - < d$$

which is an explicit Δ_0-version of the axiom (*Fullness*) in 3.6.3.
The theory CZF^ω is the extension of $CZF^{\omega-}$ by the fullness functional F together with the fullness axiom (F).
CZF^ω is thus an extension of CZF as well as of $T_\in^+$. In particular, by proposition 3.6.4,

$$CZF^\omega \vdash (\textit{Subset Collection})$$

The proof of proposition 3.6.6 immediately yields an exponentiation functional exp in terms of F via

$$\mathsf{exp}\, ab = \{ f \in \mathsf{F}a(a \times b) \mid f : a \to b \} \ = \ {}^a b$$

3.6.8 Corollary to the extended $\wedge$-interpretation theorem 3.3.19

$$CZF^\omega + \{A \leftrightarrow A^\wedge\} \overset{\wedge}{\hookrightarrow} T_\in^+$$

$CZF^\omega + \{A \leftrightarrow A^\wedge\}$ *is a conservative extension of* $T_\in^+$.

Burr [6] shows that this result does not extend to $CZF^\omega + (E_2)$ (cf. 0.1.14.4). He adapts Howard's [23] concept of *majorizability* to separate the constructive set functionals, including the fullness functional, from functionals $\wedge$-interpreting (E_2), and proves:
(*1) (E_2) is not $\wedge$-interpretable by *majorizable* functionals;
(*2) any constructive set functional of $T_\in^+$ is *majorizable*.

In order to be able to take over Howard's definition of *majorization* "literally", based on the notion of *transitive set* from definition 3.1.15, Burr uses the concept of *transitive closure* of a set and its associated partial order relations $<$ and $\leq$ which share important properties with the $\in$-relation:

3.6.9 Definition of transitive closure TC and induced partial orders $<$ and $\leq$

The **transitive closure** TCt of a set t is defined by $\in$-recursion:

$$TCt = t \cup \bigcup \{ TCx \mid x \in t \}$$

It gives rise to relations $<$ and $\leq$ on sets:

$(<) \qquad s < t :\Leftrightarrow s \in TCt$

$(\leq) \qquad s \leq t :\Leftrightarrow t \geq s :\Leftrightarrow s \in TC\{t\} \Leftrightarrow s \in TCt \lor s = t$

Elementary properties of TC and the relations $<$ and $\leq$ are:

3.6.10 Lemma

(1) $T_\in$ *proves that* TCt *is the smallest transitive set containing* t:

$$T_\in \vdash tran(TCt) \land t \subset TCt \land (tran(s) \land t \subset s \to TCt \subset s)$$

(2) $T_\in$ *proves:* $<$ *is transitive and* $\leq$ *is reflexive and transitive.*

Proof of (1). Clearly, $t \subset TCt$. To show the transitivity of TCt, let $s \in TCt$. Either $s \in t$ whence $s \subset TCs \subset TCt$; or $\exists x \in t\ s \in TCx$. Then by I.H. $s \subset TCx \subset TCt$, and by $\in$-induction, $s \subset TCt$ follows.
Finally, assume $tran(s) \land t \subset s$. For all $y \in TCt$, either $y \in t$ and hence $y \in s$, or $\exists x \in t\ y \in TCx$. Then $x \in s$, because s is transitive, and by I.H., $y \in s$. By $\in$-induction, $TCt \subset s$ follows.
Proof of (2). If $r \in TCs$ and $s \in TCt$, then, by transitivity of TCt, $s \subset TCt$ and by (1), also $TCs \subset TCt$. Hence $r \in TCt$. The rest is standard.

In contrast to the $\in$-relation, the $<$-relation is transitive, and yet, like the $\in$-relation, the $<$-relation satisfies the principles of transfinite induction and recursion:

3.6.11 Proposition: Induction along transitive closure

$T_\in$ *is closed under the rule*

$(TC\ IND) \qquad \forall x \in TCy\ F[x] \to F[y] \vdash F[t]$

Proof. By the premise of $(TC\ IND)$, $\forall y \in z \forall x \in TCy\ F[x]$ implies

$$\forall y \in z\ F[y] \land \forall y \in z \forall x \in TCy\ F[x] \equiv \forall x \in TCz\ F[x]$$

From this implication, $(T\ IND)$ concludes $\forall y \in TC\{t\}\ F[y]$ which, because of $t \in TC\{t\}$, implies $F[t]$.

3.6.12 Proposition: Recursion along transitive closure

For any type τ, *there is a functional*
$$\mathsf{R}_\tau^{TC} : ((o \to \tau) \to o \to \tau) \to o \to \tau \ in\ T_\in$$
for which $T_\in$ *proves for terms* $g : (o \to \tau) \to o \to \tau$ *and* $t : o$
$(R^{TC}) \qquad \mathsf{R}_\tau^{TC} gt = g(\mathsf{R}_\tau^{TC} g \restriction TCt)t$

Proof. Given g and y^o, we simultaneously by proposition 3.1.13 and by case distinction 3.1.6 define terms f and $f'y$, both of type $o \to \tau$, as follows:

(1) $\qquad\qquad f'yx = fx$ if $x \in TCy$

which amounts to

(2) $\qquad\qquad f'y \restriction TCy = f \restriction TCy$ if $\{x\} \cap TCy \neq 0$,

(3) $\qquad\qquad f'yx = g(f'y \restriction x)y$ if $x \notin TCy$ i.e. if $\{x\} \cap TCy = 0$, and

(4) $\qquad\qquad fy = f'y(TCy)$

Since $TCy \notin TCy$, (4), (3) and (2) imply

$$fy = f'y(TCy) = g(f'y \restriction TCy)y = g(f \restriction TCy)y$$

so that $\mathsf{R}^{TC} := \lambda g.f[g]$ satisfies (R^{TC}).

It is seen by $(TC\ IND)$ and case distinction 3.1.6 that f and f' are totally defined: By I.H., let fx be defined for all $x \in TCy$. Then $f'yx$ is defined for all $x \in TCy$ by (1) whence (2) holds. Then $f'yx$ is defined for all $x \notin TCy$ by (3), in particular for $x = TCy$. By 3.1.6, $f'yx$ is defined for all x, and fy is defined by (4). By $(TC\ IND)$, fy and $f'y$ are then defined for all y.

Burr [6] considers a special case $\restriction^+$ of restriction to transitice closure and a recursor R^+ using $\restriction^+$ as restriction operator which is useful in constructing a functional majorizing the $\in$-recursor R , as will be seen below.

3.6.13 Lemma: +-recursion

For the special restriction $\restriction^+\colon (o \to \tau) \to o \to o \to \tau$, *defined by*

$$(f \restriction^+ t)xz = \{fyz \mid y < t\}$$

(here, z is a tuple of variables of type tuple σ, where $\tau = \sigma \to o$) there is a recursor R^+ in $T_\in$ satisfying

$$R^+gt = g(R^+g \restriction^+ t)t$$

Proof. By its definition, $(f \restriction^+ t)xz = \{(f \restriction TCt)yz \mid y < t\}$ i.e.

$$f \restriction^+ t = \mathsf{K}\lambda z.\{(f \restriction TCt)yz \mid y < t\}$$

For $f = R^+g$, $R^+gt = g(R^+g \restriction^+ t)$ yields

(5) $\qquad\qquad R^+gt = g(\mathsf{K}\lambda z.\{(R^+g \restriction TCt)yz \mid y < t\})t$

showing that R^+ is definable by TC-recursion 3.6.12. To be explicit, put $hwt = g(\mathsf{K}\lambda z.\{wyz \mid y < t\})t$. Then

$$\mathsf{R}^{TC}ht = h(\mathsf{R}^{TC} \restriction TCt)t = g(\mathsf{K}\lambda z.\{(\mathsf{R}^{TC}h \restriction TCt)yz \mid y < t\})t$$

is (5) for $R^+g := \mathsf{R}^{TC}h$.

It will be noticed that $(f \restriction^+ t)x$ is independent of x. $\restriction^+$ may therefore also be viewed as a kind of *supremum operator*.

With the relation $\geq$ as defined above, Burr [6] can literally take over the concept of *majorization* from Howard as presented in section 1.10 in order to separate the functionals of $T_{\in}^{+}$ from functionals $\wedge$-interpreting type-extensionality. We give the definition again in detail – now with reference to the relation $\geq$ as introduced in 3.6.9.

3.6.14 *Recursive definition of the relation f^* majorizes f, f^* maj f for $f^*, f : \tau$ by recursion on linear types τ.*

1. For $f^*, f : o \quad$, f^* *maj* f iff $f^* \geq f$.
2. For $f^*, f : \sigma \to \tau$, f^* *maj* f iff, for all g^*, g of type σ,
   $$g^* \ maj \ g \ \text{implies} \ f^* g^* \ maj \ fg.$$
$f : \tau$ is **majorizable**, if there is $f^* : \tau$ such that f^* *maj* f.

Also the tuple notation for *maj* is the same as in 1.10.5, and lemma 1.10.6 holds for set theoretic functionals by the same argument as it does for arithmetic ones:

3.6.15 *Lemma*

Let τ, ρ be linear types and let σ be a type tuple of length n such that $\tau = \sigma \to \rho$, and let $f^*, f : \tau$.
(1) f^* maj f iff for all $g^*, g : \sigma$ g^* maj g implies $f^* g^*$ maj fg.
In particular:
(2) f^* maj f and g^* maj g implies $f^* g^*$ maj fg,
and in case $\rho = o$,
(3) f^* maj f iff for all $g^*, g : \sigma$ g^* maj g implies $f^* g^* \geq fg$.

Also lemma 1.10.7 holds in $T_{\in}$, if the maximum is replaced by the transitive closure:

3.6.16 *Lemma*

Let $f', f : o \to \sigma \to o$, σ a type tuple. If $f't$ maj ft for all $t : o$, then the term f^* given by

$$f^* t z = \{ f' x z \mid x \leq t \}$$

with z a tuple of variables of type tuple σ, majorizes f.

Proof. Let $t^* \geq t$ and $b^* \; maj \; b$. Then by 3.6.15 (3),

$$f^* t^* b^* = \{f'xb^* \mid x \leq t^*\} \geq f'tb^* \geq ftb$$

and $f^* \; maj \; f$.

This lemma together with the following one allows the construction of a functional R^* in $T_\in$ majorizing R.

3.6.17 *Lemma*

If $g^* \; maj \; g$ *, then* $R^+ g^* t \; maj \; \mathsf{R}gt$ *for all* $t : o$.

Proof by $(TC \; IND)$. Let by I.H. $R^+ g^* x \; maj \; \mathsf{R}gx$ for all $x \in TCt$, i.e.

$$R^+ g^* x z^* \geq \mathsf{R}gxz \quad \text{for} \quad x < t, \; z^* \; maj \; z$$

where z^*, z are tuples of variables of type tuple σ with $\tau = \sigma \to o$. By lemma 3.6.13, that implies

$$(R^+ g^* \upharpoonright^+ t)x^* z^* = \{R^+ g^* y z^* \mid y < t\} \geq (\mathsf{R}g \upharpoonright t)xz$$

for $g^*, x^*, z^* \; maj \; g, x, z$ in case $x \in t$ as well as in case $x \notin t$, because, in the latter case, the right hand side is $= 0$. Therefore

$$R^+ g^* \upharpoonright^+ t \; maj \; \mathsf{R}g \upharpoonright t$$

and since $g^* \; maj \; g$ and $t \; maj \; t$, also

$$R^+ g^* t = g^* (R^+ g^* \upharpoonright^+ t)t \; maj \; g(\mathsf{R}g \upharpoonright t)t = \mathsf{R}gt$$

By $(TC \; IND)$, the lemma follows.

We are now in a position to construct, for all functionals of $T_\in^+$, majorizing functionals also coming from $T_\in^+$.

3.6.18 *Theorem: Majorizability of constructive set functionals within* $T_\in^+$

Any functional f in $T_\in^+$ is majorized by a functional f^* in $T_\in^+$.

Proof. 1. If $f : \sigma \to \tau$ and $g : \sigma$ are majorized by functionals f^*, g^* respectively, then fg is majorized by $f^* g^*$ by lemma 3.6.15 (2). It therefore suffices to construct majorizing functionals for the constants of $T_\in^+$.
2. $0, \omega, \mathsf{K}, \mathsf{S}$ are self-majorizing (cf. theorem 1.10.8).
3. Let f be a functional of type $o^n \to o$ (o^n standing for an n-tuple of $o's$) and let x, t, t^* be n-tuples of variables respectively functionals of type o. Then, for $t \leq t^*$, $ft \in \{fx \mid x \leq t^*\}$. Therefore, if f^* is defined by

$$f^* t = \{fx \mid x \leq t\},$$

$t \leq t^*$ implies $ft \leq f^* t^*$, and f^* *maj* f.

So, functionals Suc, Int, Imp, and F are majorized by functionals Suc*, Int*, Imp*, and F* thus defined.

4. In Uft, the argument f is of type 1; therefore, one cannot argue as above, and this is the point where the fullness functional, rather its consequence, the exponentiation axiom 3.6.5 and the exponentiation functional come into play:

$Uft = \bigcup \{fz \mid z \in t\}$ depends only on the (set) function (cf. 3.6.2)

$$g[f,t] := \{\langle z, fz \rangle \mid z \in t\} \in \exp t\{fz \mid z \in t\}$$

with $\bigcup im\ g[f,t] = \bigcup \{fz \mid z \in t\} = Uft$. So we define

$$U^* ft := \{\bigcup im\ g \mid g \in \exp xy, x \leq t, y \leq ft\}$$

If f^* *maj* $f, t^* \geq t$ and hence also $f^* t^* \geq ft$, then

$$Uft = \bigcup im\ g[f,t] \in U^* f^* t^*$$

which implies $U^* f^* t^* \geq Uft$, so that U^* *maj* U, as was to be shown.

5. We come to the $\in$-recursor R. Let g^*, g with g^* *maj* g be given. Then, for any $t : o$,

$$R^+ g^* t\ maj\ Rgt$$

by lemma 3.6.17. Putting $f'[g^*] := R^+ g^*$ and $f := Rg$, there is, by lemma 3.6.16, an $f^*[g^*]$ and hence a functional $R^* := \lambda g. f^*[g]$ in $T_\in$ such that

$$R^* g^* = f^*[g^*]\ maj\ f = Rg$$

As this holds for any g^* *maj* g , R^* *maj* R. That completes the proof of theorem 3.6.18.

This theorem shows (slightly more than) (*2) under 3.6.8 above. To show (*1), one may follow the procedure in section 1.10 quite closely. The $\wedge$-translation of (E_2) differs from the one in the arithmetic case only by replacing one occurrence of $<$ by $\in$:

3.6.19 $\wedge$-*translation of* E_2

$$(E_2)^\wedge \equiv \exists X \forall Y uv\ (\forall x \in XY uv\ (ux = vx) \rightarrow Yu = Yv)$$

To show (*1), it has to be shown that any functional X satisfying this formula is not majorizable.

3.6.20   Definition: Bounded classes

For $\sigma = \rho \to o$ with ρ a type tuple, and s a functional of type o, put

$$M(\sigma; s) := \{b : \sigma \mid \forall z^\rho \ bz \le s\} = \{b : \sigma \mid \lambda z^\rho.s \ maj \ b\}$$

If σ is a type tuple $\sigma_1, ..., \sigma_r$, then put

$$M(\sigma; s) := M(\sigma_1; s) \times ... \times M(\sigma_r; s)$$

3.6.21   Lemma

Let σ be a type tuple and s a functional of type o. If $a : \sigma \to o$ is majorizable, then there is a functional $m : o$ such that $ab \le m$ for all $b \in M(\sigma; s)$.

Proof. Let $\sigma = \sigma_1, ..., \sigma_r$ and each $\sigma_i = \rho_i \to o$ (with ρ_i a type-tuple). Then, for $i = 1, ..., r$, any $b_i \in M(\sigma_i; s)$ is majorized by $\lambda z_i^{\rho_i}.s$. If $a : \sigma \to o$ is majorized by, say, a^*, then, by lemma 3.6.15 (3),

$$m := a^*(\lambda z_1.s)...(\lambda z_r.s) \ge ab_1...b_r \equiv ab$$

3.6.22   Theorem

Functionals X satisfying $(E_2)^\wedge$ are not majorizable.

Proof. Define the following families of functionals
$\lambda t.u_t : o \to 1$ and $\lambda t.Y_t : o \to 2$ by case distinction such that
$u_t x = 0$ for $x < t$ and $u_t x = 1 := \{0\}$ for $x \not< t$;
$Y_t u = 1$ if $\forall x < t \ ux = 0$ and $ut = 1$; otherwise $Y_t u = 0$.
Clearly, for all t, $u_t \in M(1; 1)$ and $Y_t \in M(2; 1)$.

Assume X to satisfy $(E_2)^\wedge$. Then X will in particular satisfy the matrix of $(E_2)^\wedge$ for

$$Y = Y_t, \ u = u_t, \ v = 0^1 \ \text{ for all } \ t : o$$

and X will a fortiori satisfy

$$\forall t(\forall x < XY_t u_t 0^1 \ u_t x = 0 \ \to \ Y_t u_t = Y_t 0^1 = 0)$$

But $Y_t u_t = 1$. Therefore $\forall t \exists x < XY_t u_t 0^1 \quad u_t x = 1$.
However, $u_t x = 1$ holds only for $x \not< t$. Therefore finally

$$\forall t \ XY_t u_t 0^1 \not< t$$

and, since all Y_t are in $M(2; 1)$, all u_t in $M(1; 1)$ (and 0^1 in $M(1; 0)$), by lemma 3.6.21, X is not majorizable.
So also (*1) under 3.6.8 is shown. (*1) and (*2), i.e. the last two theorems together yield

3.6.23   Theorem: Non-interpretability of $(t-ext)$ and (E_2).

The axiom of extensionality $(t-ext)$ and its special case (E_2) are not $\wedge$-interpretable in $T_{\in}^+$ nor in its classical version $T_{\in}^{c+}$.

3.6.24   Corollary: Underivability of $(t-ext)$ and (E_2)

The axiom of extensionality $(t-ext)$, even its special case (E_2), is underivable in theories $\wedge$-interpreted in $T_{\in}^+$ or in $T_{\in}^{c+}$, as there are

$$CZF^\omega + \{A \leftrightarrow A^\wedge\} \quad and \quad KP\omega^\omega + (qf - ARC)$$

This follows from the last theorem by theorem 3.3.19 with corollary 3.6.8 and by theorem 3.2.16.

Bibliography

[1] Aczel, P.H.G.: The type theoretic interpretation of constructive set theory, in: A. McIntyre, L. Pacholski, J. Paris (Eds.), Logic Colloquium '77, North Holland, Amsterdam 1978, 55 - 66.

[2] Aczel, P.H.G. and M. Rathjen: Notes on Constructive Set Theory, Mittag-Leffler Technical Report No. 40, 2000/2001 (2001).

[3] Avigad, J., and S. Feferman: Gödel's functional ('Dialectica') interpretation, in: S. Buss (Ed.), Handbook of Proof Theory, Studies in Logic and the Foundations of Mathematics, Vol. 137, Elsevier, Amsterdam 1998, 337 - 406.

[4] Barwise, J.: Admissible Sets and Structures, Springer-Verlag, Berlin Heidelberg New York 1975.

[5] Brouwer, L. E. J.: Zur Begründung der intuitionistischen Mathematik III, Math. Ann. 96 (1927) 451 - 488.

[6] Burr, W.: Functional interpretation of Aczel's constructive set theory, APAL 104 (2000) 31 - 73.

[7] Burr, W.: A Diller-Nahm-style functional interpretation of $KP\omega$, Arch. Math. Logic 39 (2000) 599 - 604.

[8] Burr, W.: Concepts and aims of functional interpretations: Towards a functional interpretation of constructive set theory, Synthese 133 (2002) 257 - 274.

[9] Burr, W. und V. Hartung: A characterization of Σ_1-definable functions of $KP\omega + (uniform AC)$, Arch. Math. Logic 37 (1998) 199 - 214.

[10] Diller, J.: Zur Berechenbarkeit primitiv-rekursiver Funktionale endlicher Typen. Contributions to Mathematical Logic, ed. Schmidt, Schütte, Thiele. Amsterdam 1968, 109 - 120.

[11] Diller, J.: Functional interpretations of Heyting's arithmetic in all finite types, Nieuw Archief voor Wiskunde (3) 27 (1979) 70 - 97.

[12] Diller, J.: Logical problems of functional interpretations, APAL 114 (2002) 27 - 42.

[13] Diller, J.: Functional interpretations of constructive set theory in all finite types. Dialectica 62 (2008) 149 - 177.

[14] Diller, J.: Functional interpretations of classical systems, in: R. Schindler (Ed.), Ways of Proof Theory, Ontos Mathematical Logic, vol. 2, Ontos Verlag,

Heusenstamm 2010, 241 - 255.

[15] Diller, J., and W. Nahm: Eine Variante zur Dialectica-Interpretation der Heyting-Arithmetik endlicher Typen, Arch. Math. Logik Grundl. 16 (1974) 49 - 66.

[16] Diller, J., and K. Schütte: Simultane Rekursionen in der Theorie der Funktionale endlicher Typen. Archiv f. math. Logik u. Grundlagenf. 14 (1971), 69 - 74.

[17] Diller, J., and A. S. Troelstra: Realizability and intuitionistic logic. Synthese 60 (1984) 253 - 282.

[18] Diller, J., and H. Vogel: Intensionale Funktionalinterpretation der Analysis, in: J. Diller and G.H. Müller (Eds.), ISILC Proof Theory Symposium Kiel 1974, LNM 500, Springer 1975.

[19] Feferman, S.: Gödel's Dialectica interpretation and its two-way stretch, Computational Logic and Proof Theory, Springer, Berlin 1993, 23 - 40.

[20] Gaspar, J., and P. Oliva: Proof interpretations with truth, Math. Logic Quarterly 2011

[21] Gödel, K.: Über eine bisher noch nicht benützte Erweiterung des finiten Standpunktes, Dialectica 12 (1958) 280 - 287.

[22] Gödel, K.: Collected Works, vol. II, Publications 1938 - 1974, S. Feferman (Ed.), The Clarendon Press, Oxford University Press, New York 1990.

[23] Howard, W.A.: Functional interpretation of bar induction by bar recursion, Compositio Math. 20 (1968) 107 - 124.

[24] Howard, W.A.: Ordinal analysis of bar recursion of type o. Chicago 1969.

[25] Howard, W. A., and G. Kreisel: Transfinite induction and bar induction of types zero and one, and the role of continuity in intuitionistic analysis. J. Symb. Logic 31 (1966) 325 - 358.

[26] Jensen, Ronald B., and Carol Karp: Primitive recursive set functions, in: D.S. Scott (Ed.), Axiomatic set theory Vol. 13 Part 1. American Mathematical Soc., 1971, 143 - 176.

[27] Jørgensen, K.F.: Functional interpretation and the existence property, Math. Logic Quarterly 50 (2004) 573 - 576.

[28] Kleene, St. C.: On the interpretation of intuitionistic number theory. J. Symb. Logic 10 (1945) 109 - 124.

[29] Kreisel, G.: Interpretation of analysis by means of constructive functionals of finite type, in: A. Heyting (Ed.), Constructivity in Mathematics, North Holland Publ. Co., Amsterdam 1959, 101 - 128.

[30] Luckhardt, H.: Extensional Gödel functional interpretation. LNM 306, Springer 1973.

[31] Myhill, J.: Constructive set theory. J. Symb. Logic 40 (1975) 347 - 382.

[32] Rath, P.: Eine verallgemeinerte Funktionalinterpretation der Heyting-Arithmetik endlicher Typen, Ph.D. thesis, University of Münster, 1978.

[33] Rathjen, M.:A proof-theoretic characterization of the primitive recursive set functions. J. Symb. Logic 57 (1992) 954 - 969.

[34] Rathjen, M.: From the weak to the strong existence property. JSL 163 (2012) 1400 - 1418.

[35] Schulte, D.: Hybrids of the ×-translation for CZF^ω, J. Applied Logic 6

(2008) 443 - 458.

[36] Schütte, K.: Proof Theory, Grundlehren der math. Wiss., Springer, Heidelberg/New York 1977.

[37] Shoenfield, J.R.: Mathematical Logic, Addison-Wesley Publ. Comp., Reading, MA, 1967.

[38] Spector, C.: Provably recursive functionals of analysis. Proc. Symp. Pure Math., vol.5, AMS, Providence 1962, 1 - 27.

[39] Stein, M.: Eine Hybrid-Interpretation der Heyting-Arithmetik endlicher Typen, Master's thesis, University of Münster, 1974.

[40] Stein, M.: Interpretationen der Heyting-Arithmetik endlicher Typen. Arch. math. Logik 19 (1978), 175 - 189.

[41] Stein, M.: Interpretations of Heyting's arithmetic - an analysis by means of a language with set symbols. Annals of math. logic 19 (1980), 1 - 31.

[42] Stein, M.: A general theorem on existence theorems. Z. math. Logik 27 (1980), 435 - 452.

[43] Tait, W. W.: Intensional interpretations of functionals of finite type I. J. Symb. Logic 32 (1967), 198 - 212.

[44] Tait, W. W.: Normalform theorem for bar-recursive functions of finite type, in: Proc. 2nd Scandinavian Logic Symposium, Studies in Logic and the Found. of Math. 63, North Holland, Amsterdam 1971, 353 - 367.

[45] Troelstra, A.S.: Notions of realizability for intuitionistic arithmetic and intuitionistic arithmetic in all finite types, in: Proc. 2nd Scandinavian Logic Symposium, Studies in Logic and the Found. of Math. 63, North Holland, Amsterdam 1971, 369 - 405.

[46] Troelstra, A.S.: Metamathematical investigation of intuitionistic arithmetic and analysis, Lecture Notes in Mathematics 344, Springer, Heidelberg/New York 1973.

[47] Troelstra, A.S.: Introductory Note to 1958 and 1972, in: K. Gödel, Collected Works, vol. II, Publications 1938 - 1974, S. Feferman (Ed.), The Clarendon Press, Oxford University Press, New York 1990.

[48] Troelstra, A.S.: Realizability, in: S. Buss (Ed.), Handbook of Proof Theory, North Holland, Amsterdam 1998, 407 - 473.

[49] Troelstra, A.S., and D. van Dalen: Constructivism in Mathematics, Vol. II, North Holland, Amsterdam 1988.

[50] Vogel, H.: Ein starker Normalisationssatz für die bar-rekursiven Funktionale. Arch. math. Logik 18 (1976), 81 - 84.

Index